UNDERSTANDING OF ATMOSPHERIC SYSTEMS WITH EFFICIENT NUMERICAL METHODS FOR OBSERVATION AND PREDICTION

Edited by **Lei-Ming Ma, Chang-Jiang Zhang**
and **Feng Zhang**

Understanding of Atmospheric Systems with Efficient Numerical Methods for Observation and Prediction
http://dx.doi.org/10.5772/intechopen.76493
Edited by Lei-Ming Ma, Chang-Jiang Zhang and Feng Zhang

Contributors

Yan Lei, Taixia WU, Xueqi Wang, Hai Chu, Mengjuan Liu, Min Sun, Lei Chen, Pablo Campra, Lei-Ming Ma, Feng Zhang, Yining Shi, Kun Wu, Jiangnan Li, Wenwen Li, Guangqiang Zhou, Zhongqi Yu, Yixuan Gu, Luyu Chang, Chang-Jiang Zhang, Qi Luo, Yuan Chen, Juan Lu, Licheng Xue, Xiaoqin Lu, Jan Scheffel, Kristoffer Lindvall, Miao Yu, Shiguang Miao

First published in London, United Kingdom, 2019 by IntechOpen
IntechOpen is the global imprint of INTECHOPEN LIMITED, registered in England and Wales, registration number: 11086078, The Shard, 25th floor, 32 London Bridge Street
London, SE19SG – United Kingdom
Printed in Croatia

British Library Cataloguing-in-Publication Data
A catalogue record for this book is available from the British Library

Additional hard copies can be obtained from orders@intechopen.com

Understanding of Atmospheric Systems with Efficient Numerical Methods for Observation and Prediction, Edited by Lei-Ming Ma, Chang-Jiang Zhang and Feng Zhang
p. cm.
Print ISBN 978-1-83880-111-3
Online ISBN 978-1-83880-112-0
eBook (PDF) ISBN 978-1-83880-634-7

For EU product safety concerns:
IN TECH d.o.o., Prolaz Marije Krucifikse Kozulić 3, 51000 Rijeka, Croatia,
info@intechopen.com or visit our website at intechopen.com.

For EU product safety concerns:
IN TECH d.o.o., Prolaz Marije Krucifikse Kozulić 3, 51000 Rijeka, Croatia,
info@intechopen.com or visit our website at intechopen.com.

Meet the editors

Dr. Ma Lei-Ming (PhD from Nanjing University, China) is currently a professor of Shanghai Meteorological Center and also a joint professor of East China Normal University in Shanghai, China. He is also a member of Numerical Weather Prediction Committee of Chinese Meteorological Society. Dr. Ma has published on a range of topics such as numerical weather prediction, tropical cyclones, and atmospheric physics. A convective parameterization scheme proposed by Ma Lei-Ming (Ma and Tan, 2009) was introduced into the US NCAR/WRF numerical model (V3.3) and was verified to be effective in improving the false alarm problem of the original Kain-Fritsch scheme. Prof. Ma Lei-Ming has been granted with several important academic prizes, including China Excellent Young Staff Awards on Meteorological Science and Technology (2006 and 2010), China Meteorological Administration (CMA) Awards on Meteorological Science and Technology, and Shanghai Science and Technology Award, etc.

Dr. Chang-Jiang Zhang (PhD from Beijing Institute of Technology, China) is currently a professor and master supervisor of the College of Mathematics, Physics and Information Engineering, Zhejiang Normal University, China. He is the Second Level of New Century 151 Talent Project of Zhejiang Province, China. He is an outstanding class teacher of Zhejiang Normal University (2010), a middle-aged discipline leader of Zhejiang Normal University (2008), and an outstanding middle-young-aged teacher (2004). He is also a member of the Chinese Image and Graphics Society and the Chinese Meteorological Society. Dr. Zhang has published more than 80 journal and conference papers in image processing, pattern recognition, wavelet transform, multiscale geometry analysis, machine learning, and their applications in tropical cyclone center location and intensity estimation.

Dr. Feng Zhang (BS from Zhejiang Normal University; MS from Chinese Academy of Meteorological Sciences; and PhD from University of Chinese Academy of Sciences) is currently a professor in the Key Laboratory of Meteorological Disaster, Ministry of Education, Nanjing University of Information Science and Technology, Nanjing, China. Dr. Zhang's research focuses on atmospheric parameterization and its application to numerical models. He has published on a range of topics, including radiative transfer theory, cloud/radiation interaction, and climate models. His current research projects pertain to the development of a new radiative transfer theory to deal with continuous cloud microphysical property and its application in climate models.

Contents

Preface IX

Section 1 Introduction Chapter 1

Chapter 1 **Introductory Chapter: Understanding of Atmospheric Systems with Efficient Numerical Methods for Observation and Prediction 3**
Lei-Ming Ma

Section 2 Numerical Methods for Atmospheric Observation 7

Chapter 2 **Tropical Cyclone Center Determination Algorithm by Texture and Gradient of Infrared Satellite Image 9**
Chang-Jiang Zhang, Qi Luo, Yuan Chen, Juan Lu, Li-Cheng Xue and Xiao-Qin Lu

Chapter 3 **Polarization Remote Sensing for Land Observation 23**
Lei Yan, Taixia Wu and Xueqi Wang

Section 3 Numerical Weather Prediction 51

Chapter 4 **Rainfall Nowcasting by Blending of Radar Data and Numerical Weather Prediction 53**
Hai Chu, Mengjuan Liu, Min Sun and Lei Chen

Chapter 5 **Spectral Representation of Time and Physical Parameters in Numerical Weather Prediction 71**
Kristoffer Lindvall and Jan Scheffel

Chapter 6 **Atmospheric Radiative Transfer Parameterizations 87**
Feng Zhang, Yi-Ning Shi, Kun Wu, Jiangnan Li and Wenwen Li

Chapter 7 **Evaluating Cooling Tower Scheme and Mechanical Drag Coefficient Formulation in High-Resolution Regional Model 103**
Miao Yu and Shiguang Miao

Chapter 8 **Numerical Air Quality Forecast over Eastern China: Development, Uncertainty and Future 123**
Guangqiang Zhou, Zhongqi Yu, Yixuan Gu and Luyu Chang

Chapter 9 **Numerical Simulation of the Effects of Increasing Urban Albedo on Air Temperatures and Quality over Madrid City (Spain) by Coupled WRF/CMAQ Atmospheric Chemistry Model 143**
Pablo Campra

Preface

Progress in atmospheric science has always been hastened by the development of numerical methods, in particular observation, data assimilation, physical parameterization, atmospheric chemistry, etc. These technologies depend on a wide range of sciences involving diverse disciplines from mathematics and atmospheric sciences to computer sciences, which have different histories and have been developed independently. An important feature of these numerical methods is that their computing and communications infrastructures are frequently updated so as to be "state of the art." Although the technology of observation and prediction of atmospheric systems draws upon many common fields, until now the interrelatedness and interdisciplinary nature of these research fields have scarcely been discussed in one volume containing fundamental theories, numerical methods, and operational application results. This is a book to provide in-depth explorations of the numerical methods developed for a better understanding of atmospheric systems. It is edited by three editors who, with long experience in the fields of numerical weather prediction, atmospheric physics, and observation, are ideal partners for writing on the numerical methods for observation and prediction of atmospheric systems.

Prof. Lei-Ming Ma
Shanghai Central Meteorological Observatory
Shanghai Meteorological Service
Shanghai, China

Shanghai Key Laboratory of Meteorology and Health
Shanghai, China

Dr. Chang-Jiang Zhang
College of Mathematics, Physics and Information Engineering
Zhejiang Normal University
Jinhua, China

Dr. Feng Zhang
Key Laboratory of Meteorological Disaster, Ministry of Education
Nanjing University of Information Science and Technology
Nanjing, China

Section 1

Introduction Chapter

Chapter 1

Introductory Chapter: Understanding of Atmospheric Systems with Efficient Numerical Methods for Observation and Prediction

Lei-Ming Ma

Additional information is available at the end of the chapter

http://dx.doi.org/10.5772/intechopen.82468

1. Introduction

In this book, advances in numerical methods for better understanding of atmospheric systems are introduced in eight chapters, with five chapters focusing on numerical weather prediction (NWP) and three chapters on observation. By using math to model the future state of the atmosphere, NWP relies upon observations to provide both initial and verification data so as to conduct and verify a forecast. In this sense, knowing the current state of the weather by various observations is as important as developing numerical models. Remote sensing techniques, which are crucial for observation on the occurrence and development of severe weather systems over data-sparse area, are considered as some of the best choices for global observation on atmospheric and ocean systems, e.g., tropical cyclones, clouds, dust storms, etc. Thus, Chapter 1 presents an efficient algorithm for tropical cyclone (TC) center determination by using texture and gradient of infrared remote sensing image from geostationary satellite. Because remote sensing approach faces dilemma in identifying atmospheric objects from its environment, a new polarization method is presented in Chapter 2 to solve this problem. Chapters 3–8 then attempt to enhance the performance of NWP models in prediction of atmospheric systems, by using radar remote sensing data (Chapter 3), spectral method (Chapter 4), parameterization schemes (Chapters 5, 6), and atmospheric chemistry models (Chapters 7, 8), which should be valuable for reference to researchers and forecasters in the field of NWP.

2. The advances of NWP

Before going into the details of the book, we need to make a brief introduction on the background of NWP and its relation with each of the chapters. The advances of NWP in the past were based on the achievements in the development of theoretical principles of meteorology and computational mathematics. These achievements enable us today to create powerful systems to perform high-quality forecast of atmospheric systems by using atmospheric datasets in high spatial-temporal resolution. The first breakthrough in NWP started at the end of the 1940s, when Charney et al. [1] got the first numerical simulation by solving the barotropic vorticity equation over a single layer of the atmosphere. Then, in 1954, the first operational forecast was conducted in Stockholm by Rossby et al. by using barotropic equation. Later models employed more complete equations for atmospheric dynamics and thermodynamics to improve the performance of NWP models [2]. Updates to primitive equation models were persistently conducted based on detailed consideration of various important physical processes, for example, solar radiation, moisture, latent heat, sea surface temperature, and rainfall-convection feedback, etc. The particular role of each kind of these processes in prediction was usually evaluated with sensitivity experiments, which demands a huge amount of computer resources. To avoid this problem, in Chapter 4, a new scheme called Generalized Weighted Residual Method (GWRM) is proposed to efficiently identify the parameter dependency of NWP model. The physical processes identified above are usually considered by model based on parameterization schemes [3], as there always are certain scales that are unresolved by the model grids. To first order, parameterization involves the representation of a process in terms of its known relationships to dependent variables resolved on the model grid. Some common processes that are parameterized by modern NWP models include radiation, planetary boundary layer, surface layer, shallow and deep cumulus convection, and cloud microphysics. One typical example is cloud parameterization, which can infer the properties of unresolved clouds and determine whether or not they should be present. Another example is about turbulent motion. Given wind shear and stability in grid scale, a turbulent parameterization scheme can infer the properties of subgrid-scale turbulent motions. Usually, there are many ways that physical processes can be parameterized, each with assumptions, strengths, and weaknesses. It is noticed that, for any parameterized process, the parameterization is just an approximation and thus a major source for model error. Chapters 5 and 6 will introduce in detail the design of new parameterization schemes and its application in prediction of atmospheric systems.

The advances of numerical methods on physical parameterization and observation greatly contribute to the development of numerical models, including global model for prediction of large scale processes and regional models for mesoscale application. Currently there are more than 13 centers worldwide with operational global NWP capability (BoM—Australia, CMA—China, CMC—Canada, CPTEC—Brazil, DWD—Germany, ECMWF—Europe, IMD/NCMRWF—India, JMA—Japan, KMA—Republic of Korea, MF—France, NOAA/NCEP—USA, ROSHIDROMET—Russian Federation, and UKMO—UK). Most of these global NWP centers are typically running a combination of global models at grid spaces less than 30 km and high resolution regional models with grid spaces less than 12 km, according to the

Annual WMO Technical Progress Reports. Regional models, which is more convenient and flexible than global model for its relative low requirement on computational resources and datasets, are generally run for national or regional domains centered around the immediate area of the NWP center. Here, most of the regional models used in this book employed the background fields provided by one of the global models mentioned above. Among them, the Weather Research and Forecasting (WRF) model developed by National Center for Atmospheric Research (NCAR) might be one of the regional models most popularly used by the NWP community in the world. The WRF model features dynamical cores, a data assimilation system, and a software architecture allowing for parallel computation and system extensibility. In this book, Chapter 3 will incorporate radar remote sensing data into high-resolution WRF model and evaluate its performance in numerical prediction of local weather systems.

In addition to the application in weather prediction, WRF model can also be coupled with atmospheric chemistry, namely WRF-Chem, for simulation on the emission, transport, mixing, and chemical transformation of trace gases and aerosols simultaneously with the meteorology. The WRF-Chem model has been successfully employed for investigation of regional-scale air quality, field program analysis, and cloud-scale interactions. In Chapter 7, the authors will share their experience on using WRF-Chem model for air quality numerical forecast over eastern China. Similar to the use of WRF-Chem in chemical weather prediction, another numerical model, the Community Multiscale Air Quality (CMAQ) model has been developed in US since 1998, to provide fast, technically sound estimates of ozone, particulates, toxics, and acid deposition. Because it includes information about the emissions and properties of compounds and classes of compounds, CMAQ can inform users about the chemical composition of a mixture of pollutants. This is particularly useful when measurements only give insight into aggregate details, like total particulate mass. CMAQ's generalized and flexible formulation has enabled incorporation of alternate process algorithms and numerical solution methods. This has allowed inclusion of new science in the model to address increasingly complex air pollution issues. In Chapter 8, the author will use a coupled WRF/CMAQ model to simulate the effects of increasing urban albedo on air temperatures and quality over Madrid city of Spain.

3. Future work

With the increase of computer resources, the horizontal resolution of operational global numerical models reaches the gray-zone resolution at tens of km or finer, even within several 100 m. A significant focus of research over the coming years is likely to be on improving the depiction of cloud with cloud resolving models, equipped by high resolution observations of cloud and its environment. However, the implementation of cloud resolving model into operational systems is a highly complex task, and is increasingly undertaken in a collaborative framework of national centers, in collaboration with academic partners in universities and other research institutes.

Author details

Lei-Ming Ma

Address all correspondence to: malm@typhoon.org.cn

Shanghai Meteorological Service, Shanghai, P.R. China

References

[1] Charney J, Ragnar F, von Neumann J. Numerical integration of the barotropic vorticity equation. Tellus. 1950;**2**(4):237-254

[2] Kalnay E. Atmospheric Modeling, Data Assimilation and Predictability. New York, USA: Cambridge University Press; 2003;**18**(20):1-3. ISBN: 0-521-79179-0

[3] Stensrud DJ. Parameterization Schemes: Keys to Understanding Numerical Weather Prediction Models. New York, USA: Cambridge University Press; 2007. p. 137. ISBN: 978-0-521-86540-1

Section 2

Numerical Methods for Atmospheric Observation

Chapter 2

Tropical Cyclone Center Determination Algorithm by Texture and Gradient of Infrared Satellite Image

Chang-Jiang Zhang, Qi Luo, Yuan Chen, Juan Lu,
Li-Cheng Xue and Xiao-Qin Lu

Additional information is available at the end of the chapter

http://dx.doi.org/10.5772/intechopen.79831

Abstract

A novel algorithm for tropical cyclone (TC) center determination is presented by using texture and gradient of infrared satellite image from geostationary satellite. Except those latter disappearing TC satellite images that are little valuable to a TC center determination, generally other periods of TC, all have an inner core. And the centers are generally determined in the inner core. Based on this, an efficient TC center determination algorithm is designed. First, the inner core of a TC is obtained. Then, according to the texture and gradient information of the inner core, the center location of the TC is determined. The effectiveness of the proposed TC center determination algorithm is verified by using Chinese FY-2C stationary infrared satellite image. And the location result is compared with that of the "tropical cyclone yearbook," which was compiled by Shanghai Typhoon Institute of China Meteorological Administration. Experimental results show that the proposed algorithm can provide a new technique that can automatically determine the center location for a TC based on infrared satellite image.

Keywords: tropical cyclone, infrared satellite image, center location, gradient, texture

1. Introduction

Tropical cyclone (TC) is one of the most serious natural disasters in the world. Satellite remote sensing technology can not only observe the whole situation but also the structural characteristics and center information of a TC. It is the current main method of the TC center location.

Dvorak technique (DT) is widely used to TC center location. DT, which was proposed by [1, 2] in the 1970s was the most important method for TC analysis in the world. Nowadays, based on the DT, advanced Dvorak technologies [3–5] were developed to determine the TC center location. Li et al. [6] put forward a new artificial intelligence algorithm for the eye TC center location. First, the image segmentation technique was used to extract the TC cloud features. And then, the mathematic morphology opening operation and the filling operation are used to obtain the contour of the TC eye. The gravity calculation method is applied to locate the TC center. Comparing the data to the Central Meteorological Observatory of China, the experimental results show that the deviation is smaller. Li et al. [7] proposed an improved genetic algorithm to determine the TC center location. The theoretical analysis and experimental results show that this method is reasonable and effective. Yang et al. [8] combined pattern matching with the vector field analysis to determine the TC center position. The proposed method is very efficient for the TCs, which have clear spiral structure. Fan et al. [9] used 85 GHz vertical polarization channels in specail sensor microwave/imager (SSM/I) for TC center location. According to the distributed characteristics of brightness and temperature, the TC of its band is analyzed and classified. The average gap of the best route in the test results of TC is smaller. Wei et al. [10] presented a novel spiral band model to automatically locate the TC center. And particle swarm optimization was used to optimize the model parameters. Experimental results show that this model can achieve the best average error in both latitude and longitude in comparison with the best track data. Zhang et al. [11] proposed a novel intelligent automatic system framework, which was based on the image process technology to locate the TC center. According to the symmetry shape and spiral movement feature of TC, a rotation matching methodology is used to catch the track of TC. Compared with the data of Shanghai Meteorology Center, experimental results show that this system is efficient and the performance errors are acceptable for practical applications. Zheng et al. [12] established the TC forecast service production system based on geographic information system (GIS) technology and realized the function construction. It had met the requirements of the TC center forecast service production in the Central Meteorological Observatory. Recently, some researchers used deviation angle variance [13] and pattern matching algorithm [14] for TC center detection.

Although there are some advantages in the above methods, some defects are existing. The subjectivity of DT is stronger, and the degree of automation is limited. The location effect for the smaller intensity cyclone is poorer. The method based on mathematical morphology is suitable for tropical cyclone whose morphology characteristic can be identified easily. If it is applied to the weak or atypical morphology cyclone, a big deviation will be resulted. Intelligent learning method requires a lot of experimental data and experience accumulation. It is suitable for recognizing the particular structures. Wind field analysis method is applicable to locate the tropical cyclone center, which is weak and whose circulation center is not clear. When the intensity level of tropical cyclone is higher, the center location is not accurate because it is easily affected by the resolution and heavy rain [11]. The temperature/humidity structure retrieval method is also only applied to locate the tropical cyclone center whose intensity is strong. The tempo-spatial movement matching method applied movement information of adjacent satellite images to determine the TC center position [11]. But the computation burden of the method is large. Generally, center location for eye TCs has good results. But it is not very satisfying for non-eye TCs due to some reasons. Therefore, it is very important to improve the accuracy of automatic TC center location.

Here, we extract the inner core region from an infrared satellite image, which contains a TC initially. Then, according to the characteristic of the position of TC center, which is the intersection of the ambient airflow and the different textures between the inner core region and other cloud area, we determine the TC center further. And the texture is uniform and smooth in the inner core region. The gray values of the inner core region are higher than those which are on the edge of the TC. So, we calculate the fractal dimension and 3-s order statistical parameters of a gray-gradient co-occurrence matrix to determine the inner core region of the TC. Then, according to the gradient information of the inner core region, the center location of the TC is determined.

2. Extract the inner core region

In general, the TC cloud consists of spiral cloud band, inner core region, and TC center area. If the inner core region is extracted from the TC cloud, it will reduce the area, which contains the TC center. The average gray values of the inner core region in an infrared satellite image are high and the difference between the gray values of pixels is small. The gray-gradient co-occurrence matrix (GGCOM) can be used to measure the gray value difference in an infrared satellite image. The fractal dimension is a useful to represent texture characteristics of the infrared satellite image. The inner core region of a TC is extracted by combining the fractal dimension with GGCOM.

2.1. GGCOM

The analysis method of the gray-gradient co-occurrence matrix is to extract the texture feature by the comprehensive information about the gray values and gradient of the images. Because the scales of different characteristic parameters are different, we normalize the gray values and gradient values of the image. Details for GGCOM computation refer to our previous work [15].

2.2. Fractal dimension

According to the characteristics of the satellite cloud image, we use the method by Sarkar and Chaudhuri [16] to calculate the box fractal dimension.

If the size of image is $N \times N$, we shrink the image into an value S_n by the plotting scale, $1 < S_n \leq N/2$, and S_n $(1 \leq n \leq N)$ is an integer. So, we have an estimate of the scale r, which is $r = N/S_n$.

Let an image be considered as a three-dimensional space (x, y, z). Here, (x, y) means two-dimensional coordinate, and z means the gray value of the image element. So, (x, y) plane is divided into $S_n \times S_n$ grid. There is a series of boxes, whose bottom is $S_n \times S_n$ in each grid. The height h of box can be calculated by the gray level L of the whole image, namely $[L/h] = [N/S_n]$. Each box is labeled from down to up as $1, 2, 3, \cdots$, and in which box the maximum and minimum image gray values of the (m, n) grid $(1 \leq m \leq N)$ fall is checked. If the maximum value falls into a and the minimum value falls into b, then $n_r = a - b + 1$. N_r, which is the number of the boxes that cover the entire image can be described as:

$$N_r = \sum_{m,n} n_r(m,n) \quad (1)$$

In this way, N_r is calculated by each r. r changes as the change of s. The minimum linear fit is used for $\log(N_r)$ and $\log(1/r)$, and then the slope is calculated, namely Q, which is the fractal dimension.

$$Q = \log N_r / \log(1/r) \quad (2)$$

It is necessary to consider how to select the window of subimage and r calculates the scale of fractal dimension. The window size of subimage is too small to lose the important texture or is too large to mix the edge pixels with other pixels of the image area. It affects the selection of the texture feature. Choosing different scales will also affect the correctness of the linear fitting,that is, it affects the fractal dimension Q.

2.3. Extraction of inner core region

Through a large number of experimental satellite cloud images, we select the parameters, which can reflect the characteristic of TC cloud image clearly. The specific standard for choosing is as follows: the difference of this characteristic in different TC cloud images, inner core region, and other cloud regions is the biggest. And this characteristic is not sensitive for the inner core region or other cloud region, that is, it is consistent between different samples. In this chapter, average gray scale, small gradient advantage, and gradient uniformity of the GGCOM are used as features to locate the center of the TC [15]. Then, the inner core region of TC is determined by combining GGCOM with fractal dimension.

The scales of the above four characteristic parameters vary widely because the physical significance and scope of the different characteristic parameters are different. We test 10 groups of the main cloud region of the satellite cloud images whose size is 512 × 512: (1) infrared channel 1 cloud image of the No. 0513 TC "Talim," which was obtained at 0 o'clock on August 29, 2005 (20050829_0000_IR1); (2) infrared channel 2 cloud image of the No. 0513 TC "Talim," which was obtained at 6 o'clock on August 29, 2005 (20050829_0600_IR2); (3) infrared channel 1 cloud image of the No. 0513 TC "Talim," which was obtained at 21 o'clock on August 29, 2005 (20050829_2100_IR1); (4) infrared channel 2 cloud image of the No. 0513 TC "Talim," which was obtained at 18 o'clock on August 30, 2005 (20050830_1800_IR2); (5) infrared channel 1 cloud image of the No. 0713 TC "Wipha," which was obtained at 0 o'clock on September 16, 2007 (20070916_0000_IR1); (6) infrared channel 2 cloud image of the No. 0713 TC "Wipha," which was obtained at 18 o'clock on September 16, 2007 (20070916_1800_IR2); (7) infrared channel 1 cloud image of the No. 0814 TC "Hagumi," which was obtained at 12 o'clock on September 22, 2008 (20080922_1200_IR1); (8) infrared channel 2 cloud image of the No. 0814 TC "Hagumi," which was obtained at 0 o'clock on September 24, 2008 (20080924_0000_IR2); (9) infrared channel 1 cloud image of the No. 0601 TC "Chanchu," which was obtained at 6 o'clock on May 11, 2006 (20060511_0600_IR1); and (10) infrared channel 2 cloud image of the No. 0601 TC "Chanchu," which was obtained at 12 o'clock on May 12, 2006 (20060512_1200_IR2). All the time are the universal time. About 10 groups of the satellite cloud images are shown in **Figure 1**. The corresponding main bodies of the satellite cloud images are shown in **Figure 2**. Here, the fractal dimension Q, the average gray scale $\bar{G}$, the small gradient $\bar{S}$, and the inhomogeneity of the gradient distribution $\bar{D}$ are the characteristics of the main cloud of images. The four characteristics parameter values of these 10 test cloud images are shown in **Table 1**.

From **Table 1**, we can see that the range of the fractal dimension Q is from 1 to 3; the range of $\bar{S}$ is from 0 to 1; the range of $\bar{D}$ is on the amount of 10^5; and the range of $\bar{G}$ is from 0 to 200. The difference between small and big components is large. The small component is easy to be ignored when the Euclidean distance is measured. The aforementioned four parameters should be normalized so that the same weight is used to compute the Euclidean distance.

The steps of determining the inner core region are as follows:

Step 1. Separate main body of the TC from an infrared satellite image based on our previous work [17];

Step 2. Because the size of the selection of the test cloud images is 512 × 512 or 256 × 256, we use 39 × 39 window, which is determined by the experiments to compute fractal dimension and GGCOM;

Step 3. Normalize four parameters to compute the difference value ΔD_w $(w = 1, 2, \cdots, 39)$ of the fractal dimension and GGCOM;

Step 4. Get the values of ΔD_w in each window;

Step 5. Compare with each ΔD_w, the window, which ΔD_w is the biggest is determined as the inner core region of TC.

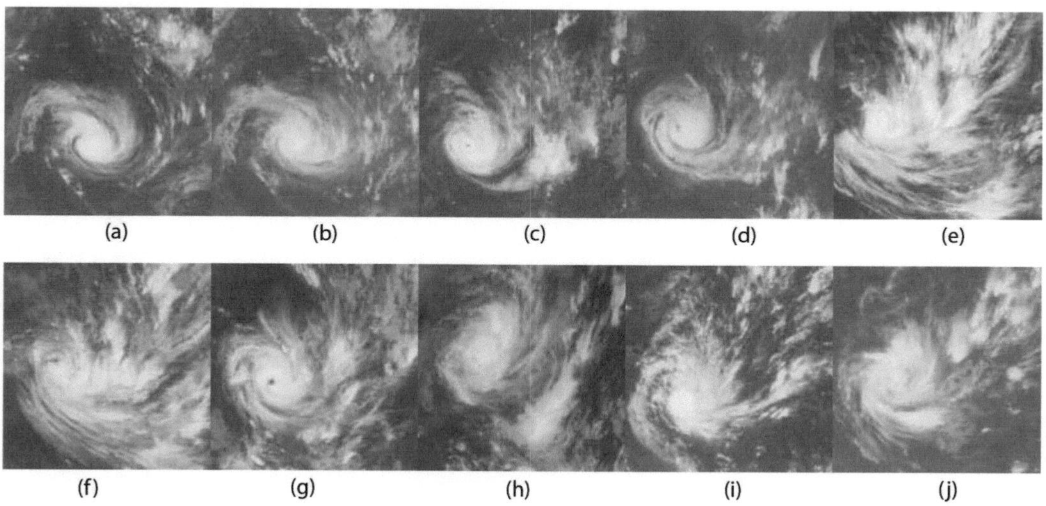

Figure 1. Ten groups of infrared channel satellite cloud images. (a) 20050829_0000_IR1; (b) 20050829_0600_IR2; (c) 20050829_2100_IR1; (d) 20070916_0000_IR1; (e) 20070916_0000_IR1; (f) 20070916_1800_IR2; (g) 20080922_1200_IR1; (h) 20080924_0000_IR2; (i) 20060511_0600_IR1; (j) 20060512_1200_IR2.

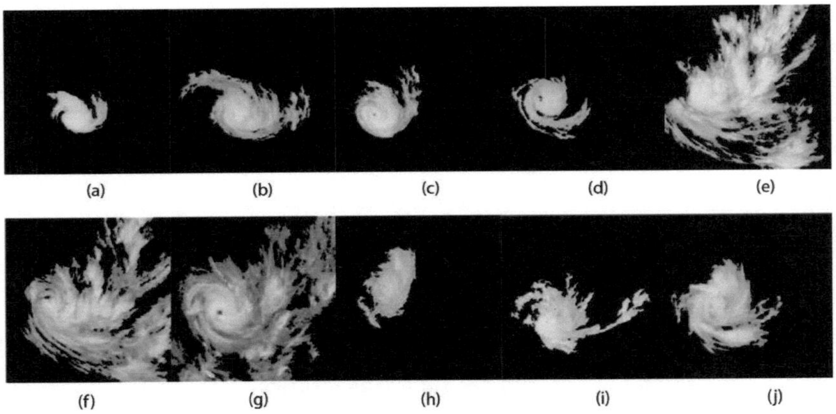

Figure 2. Corresponding main cloud of satellite cloud image. (a) 20050829_0000_IR1; (b) 20050829_0600_IR2; (c) 20050829_2100_IR1; (d) 20070916_0000_IR1; (e) 20070916_0000_IR1; (f) 20070916_1800_IR2; (g) 20080922_1200_IR1; (h) 20080924_0000_IR2; (i) 20060511_0600_IR1; (j) 20060512_1200_IR2.

Satellite cloud image	Q	$\bar{S}$	$\bar{D}$	$\bar{G}$
20050829_0000_IR1	1.2954	0.9706	2.4513×10^5	12.2469
20050829_0600_IR2	1.2189	0.9039	2.1138×10^5	31.9240
20050829_2100_IR1	1.2089	0.9414	2.3005×10^5	20.5072
20050830_1800_IR2	1.2806	0.9522	2.3546×10^5	18.2701
20070916_0000_IR1	1.1528	0.6705	1.1624×10^5	110.3207
20070916_1800_IR2	1.1633	0.6720	1.1686×10^5	100.5608
20080922_1200_IR1	1.1642	0.6264	1.0208×10^5	100.7266
20080924_0000_IR2	1.2041	0.9532	2.3573×10^5	19.4768
20060511_0600_IR1	1.1911	0.9210	2.1964×10^5	31.7581
20060512_1200_IR2	1.1729	0.8973	2.0791×10^5	40.3393

Table 1. Characteristic parameters of the main cloud in 10 groups of satellite cloud images.

Extracting the inner core region not only can reduce the region of TC center but also can eliminate the influence of the outer spirals. It can make the foundation for extracting the small gradient information. Based on these two purposes, it is not important whether to completely extract the inner core region. The key is to extract the main part of the inner core region. So a rough estimation by the size of TC cloud is feasible. The experiments also prove that the size of window can satisfy the need as per the above algorithm.

3. Determine the center of TC

It is assumed that whatever the type of the bending cloud, strong wind, eye cloud region, or type of center closed clod cloud region, the center of TC is almost located in the inner core region. After extracting the inner core region, we can determine the center of TC by the inner core region. The region near TC center has obviously the warm structure. The infrared brightness temperature gradient of the inner core region is small compared with the TC eye. The richest gradient region can be considered as the TC center region.

3.1. Compute the gradient information of inner core region

Before computing the gradient of the inner core region, Gaussian filter reduces the noise of the main body of the TC. The image and Gaussian smoothing filter are convolved as follows:

$$J(i,j) = S(i,j;\sigma^2) * I(i,j) \tag{3}$$

where $S(i,j;\sigma^2)$ represents the Gaussian function and σ^2 indicates variance. How to choose proper σ^2 is very important. $I(i,j)$ shows TC main body image.

Different variance σ^2 will be used to extract different gradient information from the main body cloud of a TC. The infrared channel 2 cloud image of the No. 0601 TC "Chanchu," which was obtained at 12 o'clock on May 12, 2006 (20060512_1200_IR2) is an example (see in **Figure 3(a)**). The image in which the variance is chosen as $\sigma^2 = 0.5$ is shown in **Figure 3(b)**. The image in which the variance is chosen as $\sigma^2 = 1.0$ is shown in **Figure 3(c)**. The image in which the variance is chosen as $\sigma^2 = 1.5$ is shown in **Figure 3(d)**. The image in which the variance is chosen as $\sigma^2 = 2.0$ is shown in **Figure 3(e)**. The image in which the variance is chosen as $\sigma^2 = 2.5$ is shown in **Figure 3(f)**.

From **Figure 3**, we can see that the influence of σ^2 for extracting the gradient information is big. When the value of σ^2 is small, the accuracy of the edge position is high. But edge details change a lot. When the value of σ^2 is small, the effect of the smooth is large and the details lose a lot.

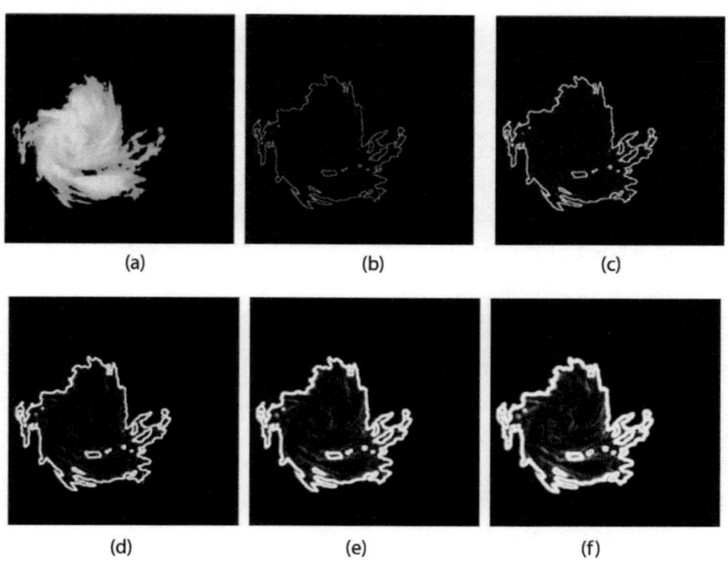

Figure 3. The influence of σ^2 for extracting the gradient information. (a) Main body cloud of TC; (b) $\sigma^2 = 0.5$; (c) $\sigma^2 = 1.0$; (d) $\sigma^2 = 1.5$; (e) $\sigma^2 = 2.0$; (f) $\sigma^2 = 2.5$.

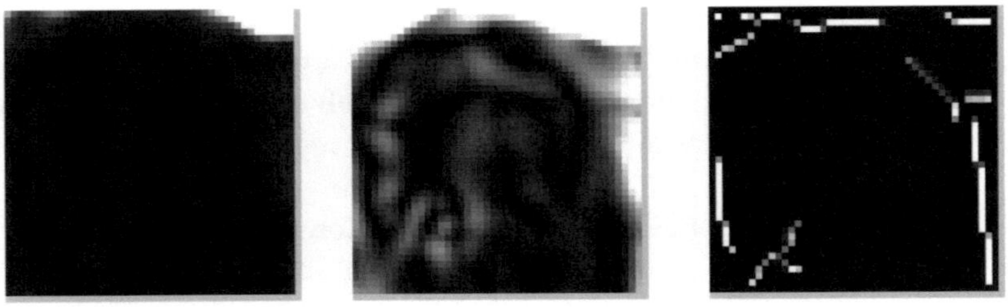

(a) (b) (c)

Figure 4. The gradient information and texture of the inner core region. (a) Gradient information when σ^2 =2.0; (b) gradient information after enhancing and (c) edge detection by Canny.

The accuracy of the edge position is low. And the gradient information is rich. Based on it, we choose the variance σ^2 is 2.0. For this condition, the gradient image of the inner core is obvious.

In order to further enhance the gradient in the inner core region. A constant 5 is multiplied to the gradient image. Canny operator is used to determine the concentrated gradient region. This region should contain the TC center. The effect of the gradient information extracted from the inner core region is shown in **Figure 4**. **Figure 4(a)** is the gradient information of the inner core region when $\sigma^2 = 2.0$. **Figure 4(b)** is the gradient information of the enhanced image. **Figure 4(c)** is the result of the edge detection by Canny operator.

We can see from **Figure 4(a)** that the gradient information of the image is not obvious. In **Figure 4(b)**, the gradient information after enhancing the image is clearer. It is convenient to do the edge detection by the Canny operator. The gradient and texture information of the inner core region is very obvious in **Figure 4(c)**.

3.2. Determine the center of TC

The region whose gradient information is the most abundant in the inner core region is the region which contains the TC center. The gradient information, which is extracted from the clear eye TC is the obvious characteristic. According to edge information of the image in Section 3.1, which is the gradient information and expresses the texture feature of the image, the region whose texture is the most abundant in the inner core region, which is called the region of interest (ROI), is the region, which contains the TC center. The geometric center of the ROI is defined as the TC center.

Generally, the average diameter for a TC eye is 45 km or so. The spatial resolution for FY-2C satellite image is 5 km. Therefore, a 9 × 9 mask is used to search the inner core region. If a closed curve is found in the inner core region, the centroid of the curve is determined as TC center. Otherwise, the region whose intersection lines are the most is chosen as the ROI.

4. Results and discussion

Infrared satellite images from Chinese geostationary satellite FY-2C are used to verify the performance of the proposed algorithm in this paper. The TC center location is used to test the

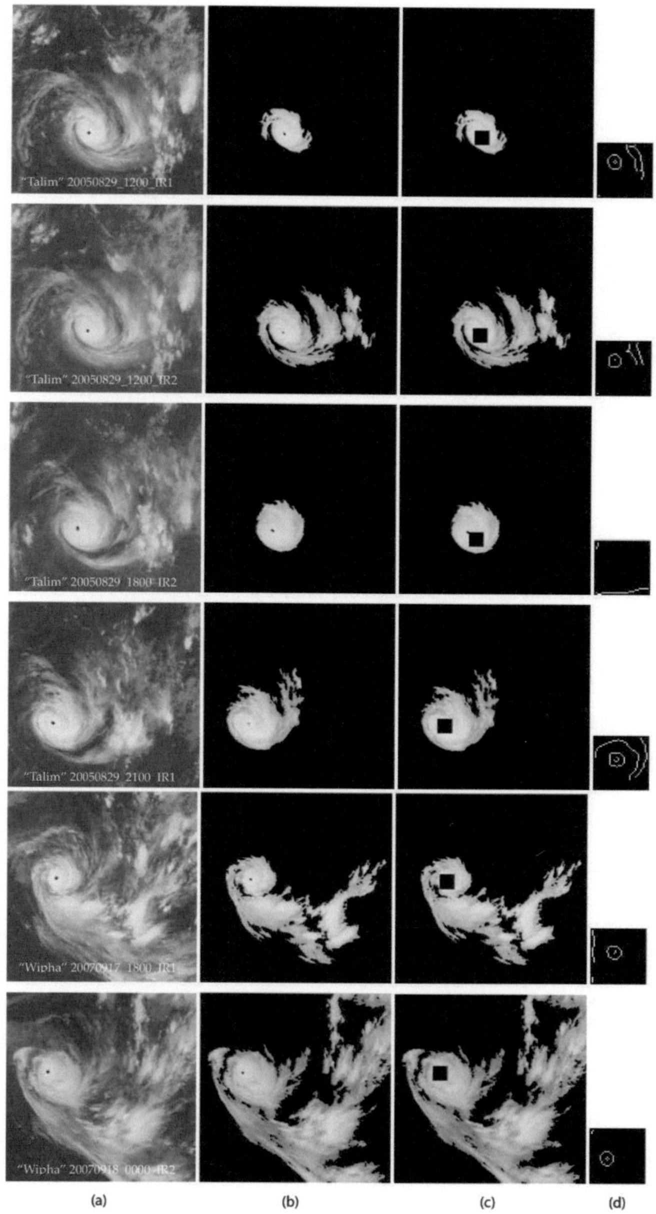

Figure 5. TC center location of eye TCs satellite images. (a) Result of the TC center location; (b) main cloud of TC; (c) position of the inner core region and (d) texture.

infrared channel 1 and the infrared channel 2 satellite images, respectively. Six groups of eye TC and six groups of non-eye TC, which are the satellite images are used to verify the proposed algorithm. All the satellite images are 512 × 512. The test satellite images of the eye TC include: (1) infrared channel 1 satellite image of the No. 0513 TC "Talim," which was obtained at 12 o'clock on August 29, 2005 (20050829_1200_IR1); (2) infrared channel 2 satellite image of the

No. 0513 TC "Talim," which was obtained at 12 o'clock on August 29, 2005 (20050829_1200_IR2); (3) infrared channel 2 satellite image of the No. 0513 TC "Talim," which was obtained at 18 o'clock on August 29, 2005 (20050829_1800_IR2); (4) infrared channel 1 satellite image of the No. 0513 TC "Talim," which was obtained at 21 o'clock on August 29, 2005 (20050829_2100_IR1); (5) infrared channel 1 satellite image of the No. 0713 TC "Wipha," which was obtained at 18 o'clock on September 17, 2007 (20070917_1800_IR1); and (6) infrared channel 2 satellite image of the No. 0713 TC "Wipha," which was obtained at 0 o'clock on September 18, 2007 (20070918_0000_IR2). The test satellite images of the non-eye TC include: (1) infrared channel 1 satellite image of the No. 0513 TC "Talim," which was obtained at 18 o'clock on August 27, 2005 (20050827_1800_IR1); (2) infrared channel 1 satellite image of the No. 0513 TC "Talim," which was obtained at 0 o'clock on August 29, 2005 (20050829_0000_IR1); (3) infrared channel 1 satellite image of the No. 0513 TC "Talim," which was obtained at 12 o'clock on August 28, 2005 (20050828_1200_IR1); (4) infrared channel 2 satellite image of the No. 0513 TC "Talim," which was obtained at 12 o'clock on August 28, 2005 (20050828_1200_IR2); (5) infrared channel 1 satellite image of the No. 0713 TC "Wipha," which was obtained at 0 o'clock on September 17, 2007 (20070917_0000_IR1); and (6) infrared channel 2 satellite image of the No. 0713 TC "Wipha," which was obtained at 0 o'clock on September 17, 2007 (20070917_ 0000_IR2). The time of these experimental images is the universal time.

4.1. Experimental group of eye TC center location

Figure 5(a)–(d) represents six eye TC satellite images and their center location results. TC center position from China Meteorological Administration is used as the reference center position. In this paper, blue "*" symbol and red "+" symbol, respectively, indicate the center position by the proposed method and reference center position.

We can see from above figures that the location results by the proposed TC center location algorithm are very close to the reference positions. Namely, the results are very close to the real center of TC. After the extraction of the main body cloud of a TC, the selected inner core region almost contains the cloud area of the eye of TC. And the texture of the inner core region is clear. The distance of the center location by the proposed algorithm and the reference position is so little. So it is difficult to identify the difference between them by the naked eye. Since

TC cloud image	Result of center location		Reference values		Error (km)
	North latitude (°)	East longitude (°)	North latitude (°)	East longitude (°)	
20050829_1200_IR1	20.95	131.89	21.00	131.90	5.66
20050829_1200_IR2	20.95	131.89	21.00	131.90	5.66
20050829_1800_IR2	21.15	130.43	21.20	130.70	30.48
20050829_2100_IR1	21.19	130.21	21.30	130.10	17.27
20070917_1800_IR1	23.93	124.67	23.90	124.60	8.45
20070918_0000_IR2	24.43	123.71	24.40	123.60	12.66

Table 2. The center location errors of the experimental eye TC group.

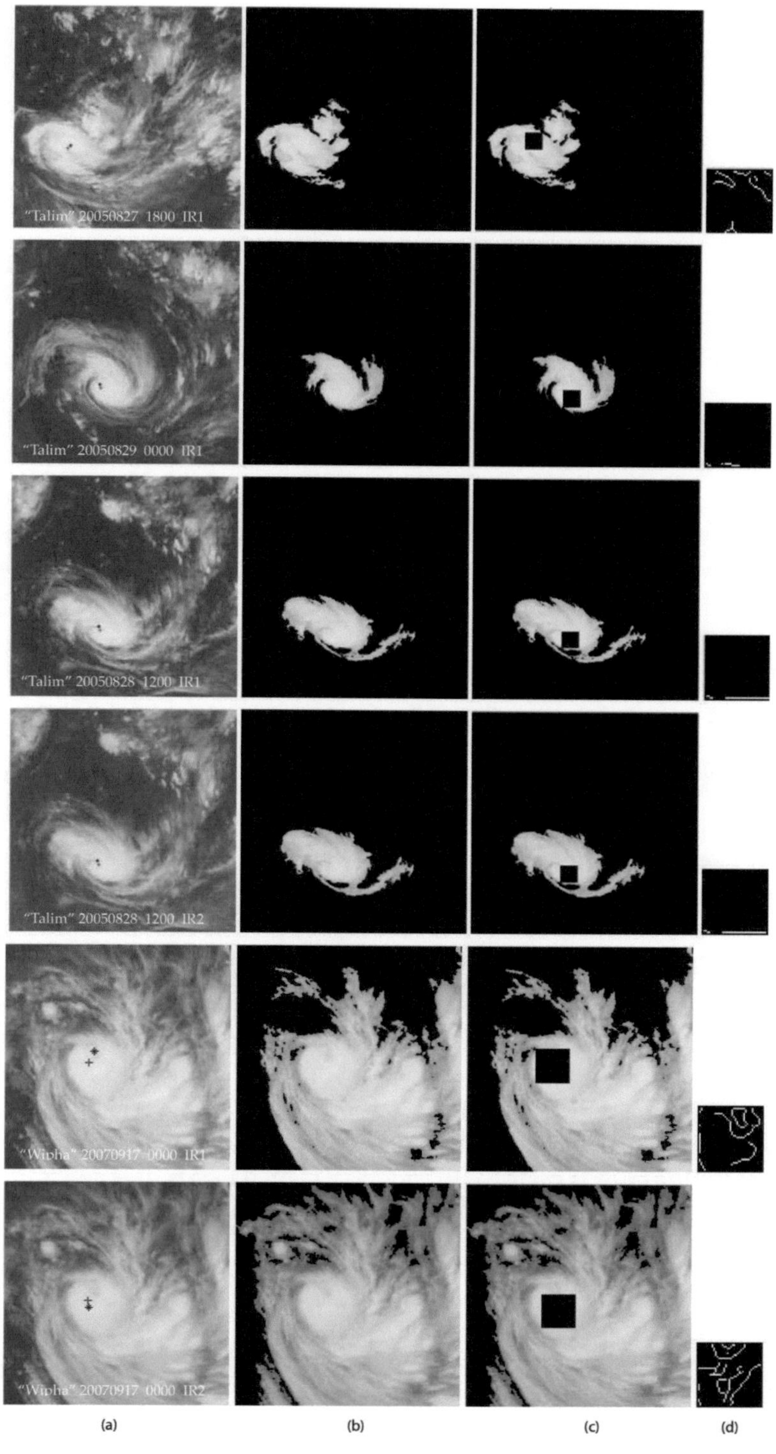

Figure 6. TC center location of non-eye TCs. (a) Result of the TC center location; (b) main cloud of TC; (c) position of the inner core region and (d) texture.

TC cloud image	Result of center location		Reference values		Error (km)
	North latitude (°)	East longitude (°)	North latitude (°)	East longitude (°)	
20050827_1800_IR1	17.59	139.43	17.90	139.90	62.50
20050829_0000_IR1	20.49	133.85	20.40	134.30	50.94
20050828_1200_IR1	20.01	136.17	19.90	136.70	60.08
20050828_1200_IR2	20.01	136.17	19.90	136.70	60.08
20070917_0000_IR1	22.07	126.91	22.40	127.70	95.03
20070917_0000_IR2	22.37	128.13	22.40	127.70	47.85

Table 3. The center location errors of the experimental non-eye TC group.

the length of 1° latitude is about 111 km and the length of 1° longitude is about 111 km all over the world, we calculate the distance error of the TC center position by the longitude and latitude errors. The center position errors of the six groups of eye TC are shown in **Table 2.**

In **Table 2,** we can find that the error of the TC center location by the proposed algorithm is below 40 km. And it is suitable for the infrared channel 1 and the infrared channel 2 cloud images.

4.2. Experimental group of non-eye TC center location

Figure 6(a)–(d) represent six non-eye TC satellite images and their center location results.

We can see from the figures that the located results by the proposed TC center location algorithm are very close to the reference positions. The effect of the non-eye TC center location is good. In order to get the more accurate location error, we calculate the distance error of the TC center position by the longitude and latitude errors. The center position errors of the six experimental groups of the non-eye TC center location are shown in **Table 3.**

In **Table 3,** we can find that the error of five groups of the non-eye TC center location by the proposed algorithm is below 70 km. Only the error of one group is more than 70 km, but it is below 100 km. In all, the proposed algorithm is suitable to be used to determine the center position for a TC by using the infrared channel 1 and the infrared channel 2 cloud images. Although the accuracy of the non-eye TC center location is slightly less than that of the eye TC center location, the error is small for the non-eye TC in which the texture is not prominent.

5. Conclusions

An effective TC center location algorithm, which is based on the fractal feature features and gradient of the infrared satellite cloud images, is proposed. The proposed algorithm is based on the assumption that the centers of a TC, which belongs to different development stages are mostly located in the inner core region. The inner core region is determined firstly. Then

http://dx.doi.org/10.5772/intechopen.79831

special gradient information and texture of the TC center are used to determine the center of TC. The proposed algorithm can be applied in the eye TC center location and the non-eye TC center location. The accuracy of the location by this proposed algorithm is higher. Moreover, the calculation of the proposed algorithm is simpler than that of the existing TC center location methods, such as the wind field analysis method, intelligent learning method based on the spiral fitting, tempo-spatial movement matching method, and so on. The proposed method is not proper to disorganized or weak TCs. Further work is needed to improve the location accuracy of TCs by using new features or techniques.

Acknowledgements

The part of work in this paper is supported by Natural Science Foundation of China (No. 41575046, No. 40805048, and No. 11026226), Project of Commonweal Technique and Application Research of Zhejiang Province of China (No. 2016C33010, No. 2012C23027), and Natural Science Foundation of Zhejiang Province of China (No. LY13D050001).

Author details

Chang-Jiang Zhang[1]*, Qi Luo[1], Yuan Chen[1], Juan Lu[1], Li-Cheng Xue[1] and Xiao-Qin Lu[2]

*Address all correspondence to: zcj74922@zjnu.edu.cn

1 Department of Electronic and Communication Engineering, College of Mathematics, Physics and Information Engineering, Zhejiang Normal University, Jinhua, China

2 Shanghai Typhoon Institute of China Meteorological Administration, Shanghai, China

References

[1] Dvorak VF. Tropical cyclone intensity analysis and forecasting from forecasting from satellite imagery. Monthly Weather Review. 1975;**103**(5):420-430. DOI: 10.1175/1520-0493(1975)103%3C0420:TCIAAF%3E2.0.CO;2

[2] Yang Hq, Yang Ym. Progress in objective position methods of tropical cyclone center using satellite remote sensing. Journal of Tropical Oceanography. 2012;**31**(2):15-27

[3] Knaff JA, Brown DP, Courtney J, Gallina GM, Beven JL. An evaluation of Dvorak technique-based tropical cyclone intensity estimates. Weather and Forecasting. 2010; **25**(5):1362-1379. DOI: 10.1175/2010WAF2222375.1

[4] Olander TL, Velden CS. The advanced Dvorak technique: continued development of an objective scheme to estimate tropical cyclone intensity using geostationary infrared satellite imagery. Weather and Forecasting. 2007;**22**(2):287-298. DOI: 10.1175/WAF975.1

[5] Velden C, Harper B, Wells F, Beven JL, Zehr R, Olander T, Mayfield M, Guard CC, Lander M, Edson R, et al. The Dvorak tropical cyclone intensity estimation technique: A satellite-based method that has endured for over 30 years. Bulletin of the American Meteorological Society. 2006;**87**(9):1195-1210. DOI: 10.1175/BAMS-87-9-1195

[6] Li H, Huang XY, Qin DY. Research the artificial intelligent algorithm for positioning of eyed typhoon with high resolution satellite image. In: Proceedings of the 2012 5th International Joint Conference on Computational Sciences and Optimization; 23-26 June 2012; Harbin. Heilongjiang: IEEE; 2012. pp. 889-891

[7] Li Y, Chen X, Fei SM, Mao KF, Zhou K. The study of a linear optimal location the typhoon center automatic from IR satellite cloud image. In: Proceedings of SPIE—The International Society for Optical Engineering; 24-24 May 2011; Beijing: SPIE; 2011. p. 81932F

[8] Yang J, Wang H. Positioning tropical cyclone center in a single satellite image using vector field analysis. Lecture Notes in Electrical Engineering. 2013;**256**:37-44. DOI: 10.1007/978-3-642-38466-0_5

[9] Fan Z. Application of satellite SSM/I data in the typhoon center location and maximum wind speed estimation [D]. Taoyuan: National Central University in Taiwan; 2004. pp. 1-2

[10] Wei K, Jing ZL, Li YX, Liu SL. Spiral band model for locating tropical cyclone centers. Pattern Recognition Letters. 2011;**32**(6):761-770. DOI: 10.1016/j.patrec. 2010. 12.011

[11] Zhang QP, Lai LL, Sun WC. Intelligent location of tropical cyclone center. In: 2005 International Conference on Machine Learning and Cybernetics, ICMLC 2005; 2005; pp. 423-428

[12] Zheng W, Wu H, Luo B, Liu Z, Lv Z, Sun L. Application of GIS to tropical cyclone product generation system. Journal of Applied Meteorological Science. 2010;**21**(2):250-256

[13] Piñeros MF, Ritchie EA, Scott Tyo J. Objective measures of tropical cyclone structure and intensity change from remotely sensed infrared image data. IEEE Transactions on Geoscience and Remote Sensing. 2008;**46**(11):3574-3580. DOI: 10.1109/TGRS.2008.2000819

[14] Jin S, Wang S, Li X, Jiao L, Zhang JA, Shen D. A salient region detection and pattern matching-based algorithm for center detection of a partially covered tropical cyclone in a SAR image. IEEE Transactions on Geoscience and Remote Sensing. 2017;**55**(1):280-291. DOI: 10.1109/TGRS.2016.2605766

[15] Changjiang Z, Yuan C, Juan L. Typhoon center location algorithm based on fractal feature and gradient of infrared satellite cloud image. In: Proceedings of SPIE-International Symposium on Optoelectronic Technology and Application; 13-15 May, 2014; Beijing. p. 92990F

[16] Chaudhuri BB, Sarkar N. Texture segmentation using fractal dimension. IEEE Transactions on Pattern Analysis and Machine Intelligence. 1995;**17**(1):72-77. DOI: 10.1109/34.368149

[17] Mengyin F, Changjiang Z, Mei J. Wavelet transform approach to segment thermal image. Journal of Beijing Institute of Technology. 2003;**12**(suppl):33-38

Chapter 3

Polarization Remote Sensing for Land Observation

Lei Yan, Taixia Wu and Xueqi Wang

Additional information is available at the end of the chapter

http://dx.doi.org/10.5772/intechopen.79937

Abstract

In the real world, vegetation, liquid surfaces, rocks, buildings, snows, clouds, fogs, etc. can all be regarded as natural polarizers. In the process of reflecting, transmitting, and scattering of electromagnetic radiations, land surface objects can produce polarized features that are related to the nature of the materials. These polarized information can determine objects' properties, and therefore, detecting the polarization information of objects becomes a new method of remote sensing. Polarization of reflected and scattered solar electromagnetic radiation adds a new dimension to the understanding of the Earth's objects' properties. The polarized bidirectional reflectance characteristics and polarized hyperspectral properties of land objects were methodically studied. The results of the polarized bidirectional reflectance characteristics can provide the theoretical basis for polarization remote sensing such as the detecting conditions, modeling and others. The polarized spectral property of the typical objects can be used as the spectral basis for polarization remote sensing. The atmospheric correction is a key problem when using polarization remote sensing method to detect land objects' information, because scattered atmospheric particles exhibit stronger polarization phenomena than land objects do. A method of using atmospheric neutral point for the separation polarization effect between objects and atmosphere has been proposed.

Keywords: polarized light, remote sensing, Earth's observation

1. Introduction

Besides properties such as intensity, frequency, and coherence, polarization is another fundamental character of electromagnetic radiation. The polarization phenomenon occurs along with the entire process of electromagnetic radiation reflection, scattering, and transmission. Since the discovery of the scattered light from the blue sky is polarized by D.F.J. Arago in 1809, the effects of various polarization parameters have been observed and measured by many scientists [1].

All of these results have clearly indicated that aerosols and natural land surfaces have large polarization effects, which, therefore, offers a powerful tool for monitoring the aerosol and land objects' properties from space. While the intensity information tells us about materials, polarization presents information about surface features, shape, shading, and roughness. Polarization information has the potential to enhance many fields of optical metrology. Just like the space applications of the polarization of microwave radiation, space polarization measurement programs are needed in the visible and near-IR spectral regions for better monitoring of the Earth's environment. Polarization of reflected and scattered solar radiation adds a new dimension to the understanding of the Earth's environmental radiation field. Furthermore, with the quick development of remote sensing, the numerical solution of the remote sensing retrieval process leads to the so-called, ill-posed inverse problem [2]. This problem is characterized by the incompleteness of the available information, the nonuniqueness of the solutions, and the noncontinuous dependence of the solutions on the input data. The polarization measurement not only can provide the intensity information of land surface objects, but also can provide some extra parameters, such as degree of polarization, polarization angle, and polarization phase information, all beneficial for the solution of the ill-posed inverse problem [3].

Currently, the research of polarization remote sensing focuses mainly on the atmosphere and climate [4, 5]. It is mainly because polarization complements the spectral and angular radiance measurements, and it produces a high sensitivity to microphysical properties of aerosol particles than do radiance measurements. Furthermore, the satellite polarization sensor POLDER's 6-km pixel size spatial resolution is a benefit for broad-scale (such as atmosphere) researches [6]. However, some researchers have found that objects on the land objects surface have strong polarized reflection. Tamalge and Curran gave a review of early attempts to use polarization information for land surface remote sensing [7]. Theoretical studies also were performed to understand the nature of polarization and to model the polarization from earth surfaces [8]. Until now, most studies were focused on atmospheric polarization than land surfaces because the polarization effect of the atmosphere is much stronger than that of land objects. Removing the atmospheric effect, therefore, would solve the bottleneck problem of using polarized information for remote sensing land surface. As a result, a major concern for the use of polarized light for the study of land surfaces is the capability to discriminate between polarization generated in the atmosphere and that generated by the surface.

2. Polarization characters of land objects

Polarization is defined as the asymmetry of vibration direction relative to spreading direction. It is a unique feature of horizontal wave. Polarization is an important feature of electromagnetic wave. Objects on the land and in the atmosphere can produce their unique polarized signals during reflection, scattering, and transmission, which means, polarization can reveal abundant information of the objects. In the nature, natural polarizers exist here and there, such as smooth leaves of a plant, soil, water surface, ice, snow, cloud, fog, etc. Reflection of sunshine by such polarizers can result in polarization. Based on this feature, polarized remote sensing provides new and potential information for objects. And polarized remote sensing has become a new Earth observation method, which is receiving more and more attention [9].

According to electromagnetic theory, light is a horizontal wave and it vibrates vertically to the transmissive direction. Based on the trajectory of light vibration, it has five polarization states: natural light (nonpolarized light), linearly polarized light, partially polarized light, round polarized light, and elliptically polarized light. Natural light has same vibration range in every direction. It may vibrate in each direction that is vertical to its spreading direction with same amplitude. If we decompose the light of all directions to only two vertical directions, then we can find the same vibration energy and amplitude in the two directions. Linear polarized light means that in the vertical plane to the spreading direction, light vector only vibrates toward a certain direction. Partial polarized light can be viewed as a mix of natural light and linearly polarized light; namely it has a vibration range in a certain direction that is superior to other directions. Round polarized light and elliptical polarized light refer to light whose vector end has a round or elliptical trajectory on the vertical plane.

Polarized light is normally embodied as elliptical polarization. We need three mutually independent parameters to describe elliptical polarization light, for example, amplitude E_x, E_y and phase difference δ (or elliptical length, short axis a, b and angle of orientation ψ). Stokes raised "Stokes Parameter" in 1852. And it became the three mostly used macromeasurable parameters. Stokes parameters not only can describe complete polarized light, but also partially polarized light.

Strokes Vector has four parameters (three of them are mutually independent). They can be shown as follows, and this set of parameters is called Strokes Vector.

$$S = [S_0 \ S_1 \ S_2 \ S_3]^T \tag{1}$$

The four Strokes parameters, which can be marked as S_0, S_1, S_2, S_3, are defined as:

$$\begin{aligned} S_0 &= \tilde{E}_x^2(t) + \tilde{E}_y^2(t) \\ S_1 &= \tilde{E}_x^2(t) - \tilde{E}_y^2(t) \\ S_2 &= 2\tilde{E}_x(t) \cdot \tilde{E}_y(t) \cos\delta \\ S_3 &= 2\tilde{E}_x(t) \cdot \tilde{E}_y(t) \sin\delta \end{aligned} \tag{2}$$

$\tilde{E}_x, \tilde{E}_y$ are components of electric vector along x, y in the selected coordinate, δ is the phase difference between two vibration components, and $\langle \ \rangle$means taking the average of time. This four-dimensional vector can embody the status of any polarized light including polarization degree. S_0 in the above equation represents polarization light intensity, S_1 represents linear polarization light component, S_2 for in 45° linear polarization light component direction, and S_3 for dextrorotation circular polarization light component.

Based on Strokes Vector, we can get polarization status information of any light as below:

$$\psi = \frac{1}{2}\arctan\frac{S_1}{S_2} \tag{3}$$

$$P = \sqrt{S_1^2 + S_2^2 + S_3^2}/S_0 \tag{4}$$

Herein, ψ is the azimuth angle of ellipse, which is also its orientation. P is a description of polarization degree of partially polarized light, and its value ranges from 0 under nonpolarized light to 1 under complete polarization. For partially polarized light, P will be the mid-value. Sometimes, I, Q, U, V are also used to replace S_0, S_1, S_2, and S_3 [10, 11].

2.1. Objects surfaces show polarized characteristic mechanism

When light is slantwise, it irradiates the land objects' surface, parts of the irradiation is reflected, and the rest is absorbed by the objects. Assuming the incidence angle is α, the refraction angle is β, and the incidence light, the reflection light and the refraction light compose in the main plane. Regardless of which vibratory direction of the incidence light is, its electric vector can be decomposed to two components: the vertical component $E_{10\perp}$ and the parallel component $E_{10=}$. Also supposing the corresponding electric vector components of the reflection light are $E'_{10\perp}$ and $E'_{10=}$, the corresponding electric vector components of the refraction light are $E_{20\perp}$ and$E_{20=}$.

When nonpolarized light reflected and refracted by the two media interfaces, the radiate directions of the reflection and refraction lights are determined by the law of reflection and refraction; however, the vibratory directions of these two lights, namely the polarization state, obey the electromagnetic theory of light. Based on the Fresnel formula, the electric vector's reflection radiant intensity of the vertical and the parallel component is as follows:

$$E'_{10\perp} = -\frac{\sin(\alpha-\beta)}{\sin(\alpha+\beta)}\cdot E_{10\perp} \tag{5}$$

$$E'_{10=} = \frac{\tan(\alpha-\beta)}{\tan(\alpha+\beta)}\cdot E_{10=} \tag{6}$$

$$\frac{E'_{10=}}{E_{10=}} = \frac{E'_{10\perp}}{E_{10\perp}}\cdot\frac{\cos(\alpha+\beta)}{\cos(\alpha-\beta)} \tag{7}$$

When $\alpha = 0^{\circ}$, because of $E_{10\perp} = E_{10=}$, so

$$E'_{10=} = E'_{10\perp} \tag{8}$$

These two components are irrelevant. After being synthesized, the reflection light is still unpolarized. Thereby, the polarization does not exist in the reflection light when the incidence light irradiates objects vertically.

When $0^{\circ} < \alpha < 90^{\circ}$,

$$|\cos(\alpha+\beta)| < \cos(\alpha-\beta) \tag{9}$$

Then, we get

$$\frac{E'_{10=}}{E_{10=}} < \frac{E'_{10\perp}}{E_{10\perp}} \tag{10}$$

From formulae (5) and (6), we found that the physical effects of the two components $E_{10\perp}$ and $E_{10=}$ are different. No matter what the polarization state of the incidence light is, the interface reflects $E_{10\perp}$ as shown in formula (5) and reflects $E_{10=}$ as shown in formula (6). Formula (10) indicates that in the reflection light, the parallel component $E'_{10=}$ is always smaller than the vertical component $E'_{10\perp}$. These two components are vector sum in the two directions that include massive polarized light of different amplitudes and different directions. Therefore, these two components are irrelevant and cannot synthesize a vector. So, the polarization state is different from the incidence light [9].

2.2. The polarized directional characteristic of land objects

The bidirectional reflectance is the common macrophenomenon of the electromagnetic wave reflection in nature. It reveals that the reflection has directivity relying on the incident direction. The ability of reflecting and dispersing the electromagnetic wave of targets are closely related to its surface structural characteristic and material composition, the surface of various targets could radiate the incident electromagnetism wave in any directions (except absorption) and form different fringing flux function of material spectrum characteristic. It has been expounded by bidirectional reflection distribution function (BRDF). In the following discussion, the observation that does not add polarized radiance is bidirectional reflection.

The multiangle polarization remote sensing intends to utilize the polarization characteristic information of the targets on ground or in air. During the process of reflecting, scattering, and transmitting the electromagnetic radiation, the multiangle polarization remote sensing can produce polarized bidirectional reflection as the remote sensing information source.

During reflection, scattering, and transmission, multiangle spectral feature and polarized feature based on intrinsic characteristics of land objects exist. By studying the multiangle spectral feature and polarized feature, their directional reflectance laws and polarized reflectance laws in 2π space can be identified. Those potential laws, together with abundant information about angle and polarization, bring new methods for remote sensing application [12].

We measured multiangle polarization reflectance and bidirectional reflectance of different types of soils, including brown forest soil, calcareous soil, clay soil, yellow soil, and humus soil with different water content using bidirectional photometer device. The light source angles and viewing zenith angles range from 0 to 60° with 10° interval, the relative azimuth between the source and the sensors (180° is the specular direction) from 0 to 360° with 10° interval. Polarizer was attached in front of the light source and can be rotated freely. The maximum polarized reflectance data were measured when the pointer was at 90°, while the minimum polarized reflectance data were measured when the azimuth was at 0°. Once the polarized data are collected, a white panel reflectance was measured immediately. Then, the ratios of the polarization and the reference data were calculated automatically. The results were stored in a database. Brown forest soil (collected from Liaoning province, China, water content is 19.7%) was taken as an example.

2.2.1. *The influence of the incidence angle*

Figures 1–3 show the 0 and 90° polarized reflectance spectra of brown forest soil at 670–690 nm spectral bands, and the incidence angles are 20, 40, and 60° respectively. In the figure, the abscissa denotes the relative azimuth angle (θ) from 90 to 270° and different designs denote different viewing zenith angles (φ) from 0 to 60°. All the data cover the 2π space of brown forest soil's surface.

In **Figure 1**, the curve of 20° incidence angle shows no obvious wave crest in the specular direction. In **Figures 2** and **3**, the curves with viewing zenith angles of 40, 50, and 60° show obvious wave peaks in the specular direction. The results indicate that with the increase of incidence angle from 10 to 60°, the spectral curves of brown forest soil in 2π space are changing gradually.

These results suggest that when the incidence angle is small, the reflection spectra of soil surface are characterized by diffuse reflection. There is almost no composite of specular and diffuse reflections. When the incidence angle increases, the reflection spectra show a specular reflection pattern. So, it is reasonable to think that there is a composite of specular and diffuse reflections. The incidence angle has influence on whether there will be a composite of specular and diffuse reflection.

2.2.2. *The influence of viewing zenith and relative azimuth angle*

Figure 1 shows when the incidence angle is small, the reflection spectra of the brown forest soil surface in 2π space have no obvious difference. The surface, thus, can be considered as a Lambert object. In **Figures 2** and **3**, when the viewing zenith angle is 40–60°, the spectral curves lost the Lambert characteristics. Because of the specular effect, there appears a wave crest in the azimuth

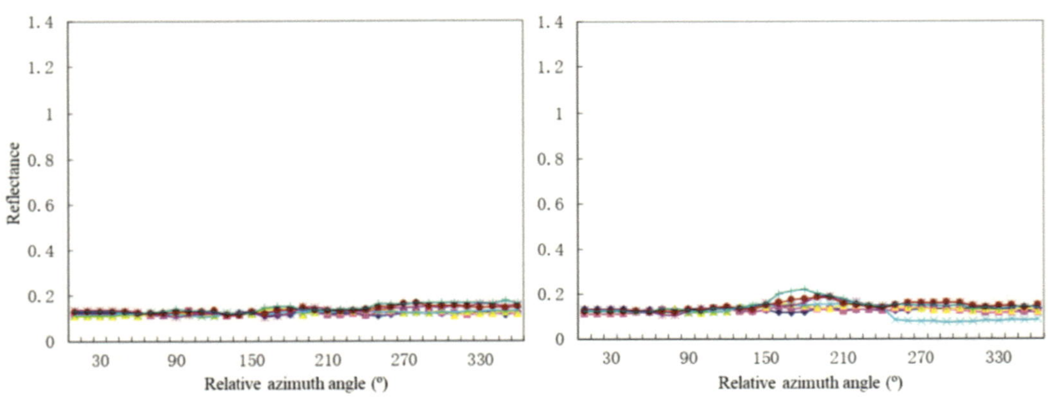

Figure 1. Polarized reflectance (0 and 90°) of brown forest soil at 670–690 nm, incidence angle = 20°.

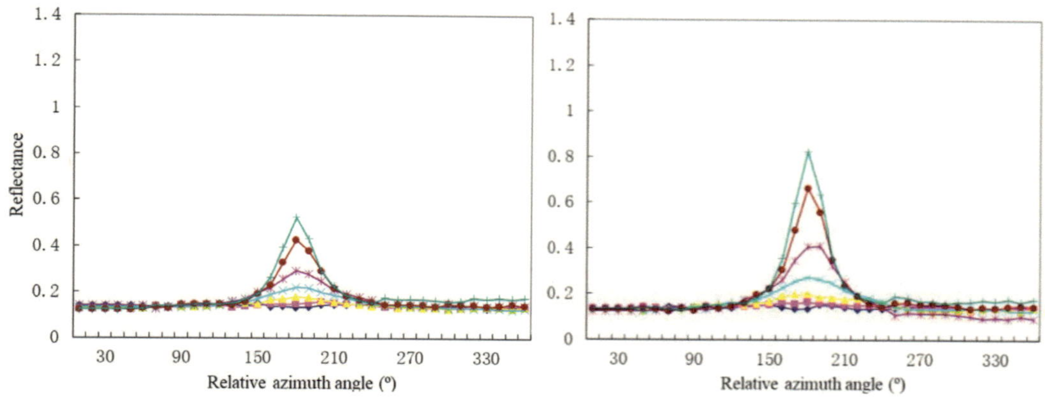

Figure 2. Polarized reflectance (0 and 90°) of brown forest soil at 670–690 nm, incidence angle = 40°.

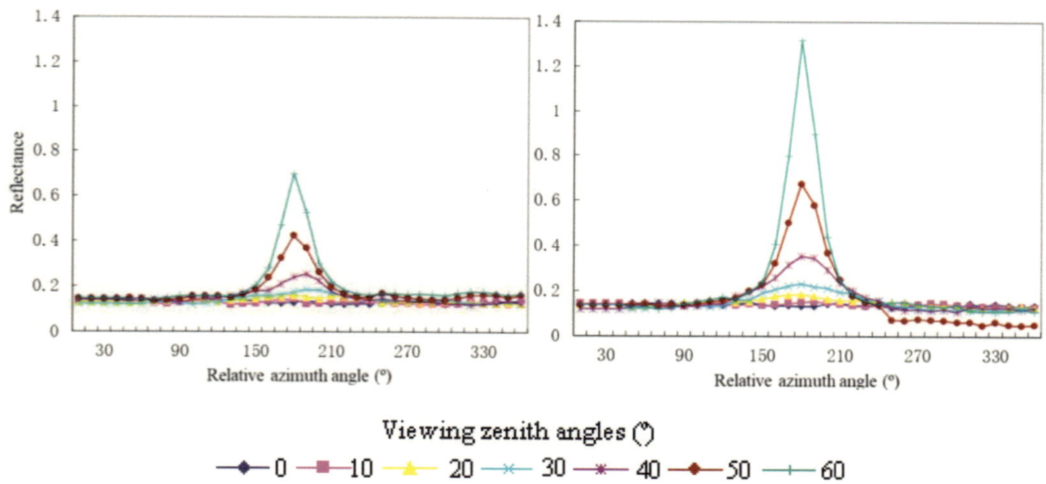

Figure 3. Polarized reflectance (0 and 90°) of brown forest soil at 670–690 nm, incidence angle = 60°.

of 135–225°. The height of the wave crest changes according to the size of viewing zenith angle. The curve for 30° has a relatively low crest, while those for 40, 50, and 60° are higher. The 180° azimuth angle is the specular direction. This phenomenon indicates that the specular effect of soil's surface is intensified when the viewing zenith angle increases, thus losing ordinary Lambert characteristics.

2.3. Polarized spectral of land objects

Polarized hyperspectral imaging is a new remote sensing method combining the benefits of polarized and hyperspectral information [13]. It has hundreds of polarized wavelengths per spatial pixel. Polarized hyperspectral imaging combines traditional two-dimensional remote sensing imaging technology and polarized spectroscopy [14, 15], allowing to obtain both images and polarized spectra of objects. The polarization measurement can not only get the intensity information of land surface objects, but can also get extra parameters, such as the degree of polarization (DoP), angel of polarization (AoP), and polarization phase information. This gives people the capability to discriminate, classify, and identify materials present in the image. Using a self-developed polarized field imaging spectrometer system (FISS-P), we collected the polarized hyperspectral images for several vegetations [16].

Ten related polarization parameters were considered in this study: *I*, *Q*, *U*, *DoLP*, *AoP*, R_0, R_{60}, R_{120}, *R*, R_1. Here, *I*, *Q*, and *U* are the three Stokes parameters; *DoLP* and *AoP* are the degree of linear polarization and angle of polarization, respectively; 0, 60, and 120° are the reflectances of three polarization azimuth angles, respectively; *R* is the nonpolarized reflectance; and R_I is the reflectance of *I*.

Figure 4 shows the *DoLP*, *AoP*, and reflectance contrast spectral curves of a *S. spectabile* leaf of the same pixel. The reflectance spectral curve of the leaf shows a typical vegetation reflectance spectral curve shape. The reflectance is low in both blue (450 nm) and red (690 nm) regions of the spectrum due to the absorption by chlorophyll in photosynthesis. It has a peak in green (540 nm) region. In near-infrared (700–800 nm) region, the reflectance is much higher than that in visible band due to the cellular structure in the leaves.

The *AoP* spectral curves of the leaf for the entire spectrum almost run parallel to the x-axis, and its values are all approximately equal to 0.5, indicating that the *AoP* spectral curves of the *S. spectabile* leaf have no unique spectral characteristics. It is not possible to use the *AoP* spectral curve as a recognition characteristic for *S. spectabile* leaves.

As for the *DoLP* spectral curves of the leaf, the reflectance is high in both blue (450 nm) and red (690 nm) regions of the spectrum. It exhibits a wave valley in green (540 nm) region. In near-infrared (700–800 nm) region, the reflectance is much lower than in visible band. The data show visible and near-infrared bands, and the *DoLP* and reflectance spectral curves of *S. spectabile* show contrasting trends.

Figure 5 displays the false color composite images of different parameters. The RGB bands of the three composite bands are the 167th band (666.8 nm), 243th band (771.3 nm), and 340th band (906.7 nm) of the FISS image. **Figure 5(a)** is the spectrum image cube of *R*; **Figure 5(b–d)** is the *I*, *Q*, *U* images, respectively; and **Figure 5(e)** and **(f)** is the DoLP and AoP images, respectively. The white area of **Figure 5(a)** and **(b)** is the white plate.

2.4. The polarization remote sensing method for water surface sun flare elimination

The sun flare produced by water mirror reflection is one of the main noises of the water color satellite images. It is an important subject to research on the sun flare and the sun flare eliminating method in remote sensing. In order to avoid the sun flare, people often set the satellite Central European Time (CET) to 12 o'clock at noon or design the sensor into multiangle scan states. However, those two methods are unable to avoid the sun flare effectively because most satellite can only broadcast vertical observation. Here, a new method for this question is presented. Combining multiangle remote sensing with polarized light, the multiangle polarized reflection method about eliminating the sun flare and the suitable time of the polarized remote sensing of

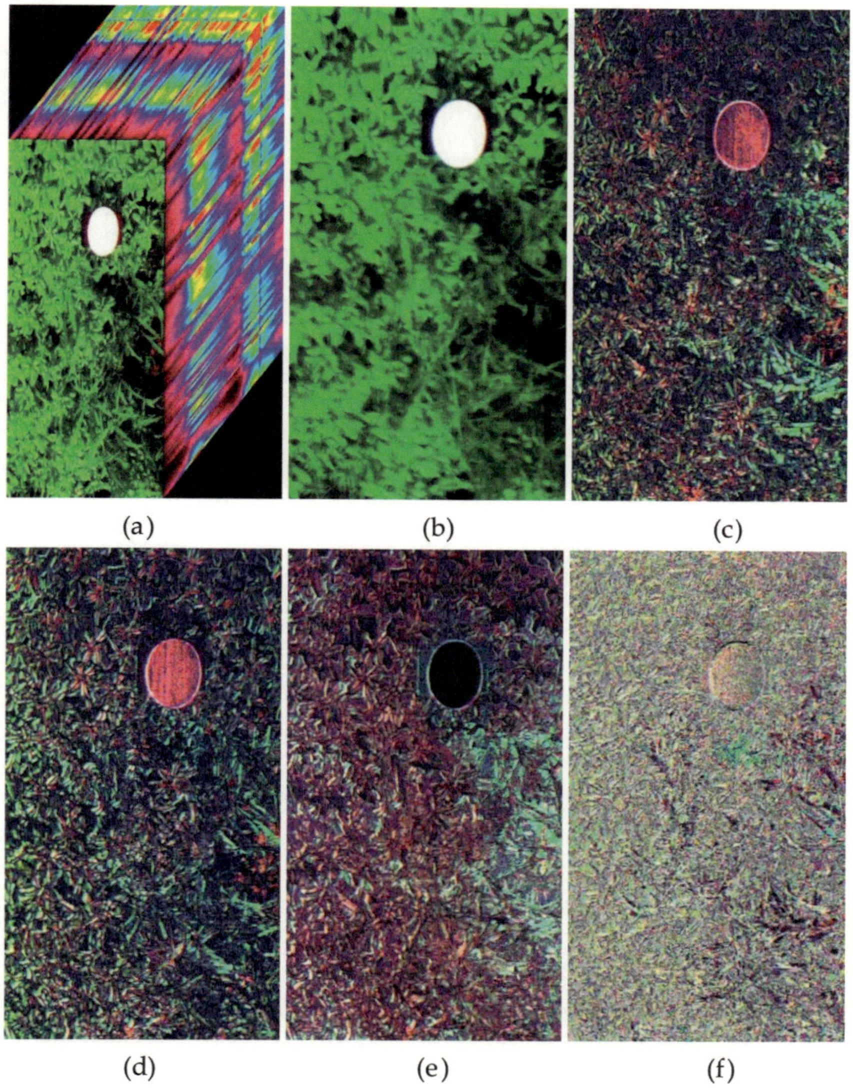

Figure 4. The *DoLP*, *AoP*, and reflectance spectral curves of a *Sedum spectabile* leaf. (a) Spectrum image cube of R, (b) I image, (c) Q image, (d) U image, (e) DoLP image, and (f) AoP image.

the water are proposed. This method will improve the utilization of the water color remote sensing images and the precision of the quantitative remote sensing.

2.4.1. Function of solar zenith and degree of polarization

When satellite CET moves from forenoon to 12 o'clock at noon, the central position of the solar flare moves from the image's east margin to the image's center; with the covering range of the solar flare changing from big to small, the distributing shape changing from long melon seeds shape to the ellipse with area gradually decreases, to a small flare at 12 o'clock at noon. When

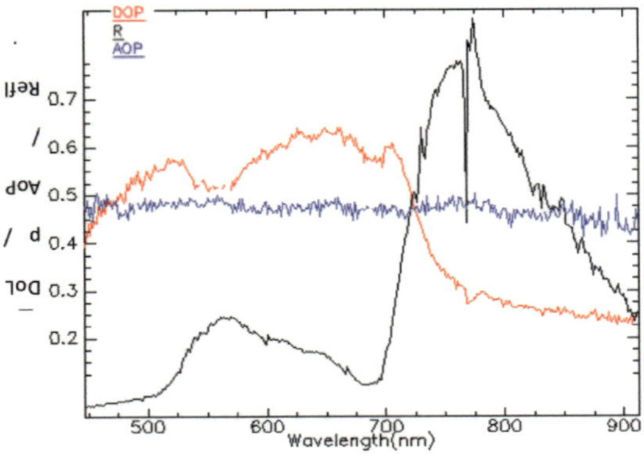

Figure 5. False-color composite images of different parameters (hyperspectral image (a), calculated Stokes parameters (*I, Q, U*) image (b, c, d), degree of linear polarization (DoLP) image (e), and angle of polarization image (f)).

the satellite CET's moving continues, the situation is just on the opposite—the solar flare's center moves toward the image's west margin, and the influence to the image by the sun gradually becomes greater. The movement of the satellite to the solar is the relative movement; so the solar's altitude angle is one of the main factors influencing the water surface solar flare's formation, size, and distributing shape.

The polarization degree is the physical quantity describing the polarized light's polarization degree, which expresses the proportion of the whole light taken by the linearly polarized light quantificationally. According to Fresnel formula, and because the light intensity is the square of the electric vector's amplitude, then

$$P = \frac{E_{10\perp}{}^{2}\frac{\sin^2(\alpha-\beta)}{\sin^2(\alpha+\beta)} - E_{10=}{}^{2}\frac{tg^2(\alpha-\beta)}{tg^2(\alpha+\beta)}}{E_{10\perp}{}^{2}\frac{\sin^2(\alpha-\beta)}{\sin^2(\alpha+\beta)} + E_{10=}{}^{2}\frac{tg^2(\alpha-\beta)}{tg^2(\alpha+\beta)}} \tag{11}$$

where $E_{\perp}$ is the component of the incident light whose electric vector is vertical to the incident interface, $E_{//}$ is the component of the incident light whose electric vector is parallel to the incident interface, α is the incidence angle, and β is the refraction angle.

When the incident light is the natural light, $E^2 = E_{=^2}$, using the refractive index to eliminate the refraction angle in Eq. (11), with the refractive index of the purified water is 1.33, yields

$$P = \frac{2\sin\alpha tg\alpha\sqrt{1.769 - \sin^2\alpha}}{1.769 - \sin^2\alpha + \sin^2\alpha tg^2\alpha} \tag{12}$$

The derivative of the P function is then calculated when $P' = 0$, $\alpha = 53.1^{\circ}$. The maximum exists at this time in the function, that is, the reflected light is completely linearly polarized light, and the incidence angle α is the Brewster angle.

The reflected light's DoP increases while the incidence angle α increases from 0° to Brewster angle gradually. When it gets to Brewster angle, the polarization degree is maximal, and then it gradually decreases. That is to say, the closer to Brewster angle, the better the reflected light's linear polarization characteristics.

2.4.2. The mechanism of sun flare elimination

The water information received by the water color remote sensing sensors mainly contains three kinds of light: (1) the light that is directly reflected by the water surface, (2) the light that arrives at the sensors through the atmospheric photon scatters, and (3) the light that is backscattered from the water body. Only the third kind of light includes water body information, it is the only source of visible light remote sensing nearly, and the first two lights constitute the background noise that must be corrected and eliminated.

According to the above discussions, light after being reflected by the water body, the polarization phenomenon exists in the reflected light, and the water body is actually the polarizer at this time. When the light's incident angle is Brewster angle of 53°, that is, the sun's altitude angle is 37°, its reflected light is totally polarized light whose electric vector is vertical to the incidence interface. By using the polarizer in front of the sensor, the polarizer's azimuth angle is adjusted and the polarization angle is made just vertical to the polarization direction of the reflected light; at this time, the reflected light totally cannot pass through the polarization sheet due to the polarizer's light-blocking effect; the information received by sensors is the atmospheric scattering and volume scattering of water body, and the intensity of the water body mirror reflection can be ignored. So, we can use the radiation transfer equation (RTE) of the atmospheric and water to reckon the water quality indicators without the solar reflected light.

The angle between the horizon and the sun is sun's altitude angle, expressed in symbol h, which can be gained by the formula

$$\sinh = \sin\varphi \sin\delta + \cos\varphi \cos\delta \cos\tau \tag{13}$$

where φ stands for the local geographical latitude, δ is the solar declination angle, and τ is solar hour angle.

According to this formula, we can calculate the sun's altitude angle in any place on the earth at any time. Then, the suitable time for the water polarization remote sensing of all the world also can be figured out when the solar altitude angle equals to 37°. Here, we only list the timetable when the sun's altitude angle equals to 37° in the place around the world on vernal equinox day (**Table 1**). The suitable time is earlier or later than 12:00 at noon. On vernal equinox day $\delta = 0°$, the time is the local time. Other time's dates can be deduced by analogy.

Table 1 shows that, on vernal equinox day, even at noon, the situation is impossible to exist that the sun's altitude angle equal to 37° in the areas exceeding south and north latitude is 53°; we cannot use the polarization remote sensing to completely eliminate the water surface mirror reflection at this time.

Combining the local solar zenith angle, in the areas of south and north latitudes between 0 and 30°, there are 12 months in 1 year that the water surface mirror reflection can be completely avoided; in the areas of south and north between 30 and 40°, there are 8 months that the water surface mirror reflection can be completely avoided; the rest may be deduced

North latitude	0°	N10°	N20°	N30°	N40°	N50°	N53°	N60°
Suitable time (h)	12 ± 3:32:00	12 ± 3:29:19	12 ± 3:20:42	12 ± 3:03:55	12 ± 2:32:53	12 ± 1:22:16	12	—
South latitude	0°	S10°	S20°	S30°	S40°	S50°	S53°	S60°

The suitable time is the time when the solar altitude angle equals to 37°.

Table 1. The suitable time for water polarization remote sensing on vernal equinox day around the world.

by analogy, and in the polar region there is no time in the whole year that the water body surface mirror reflection can be completely avoided, because the sun's altitude angle is always lower than 37°.

The above discussion is the case of completely eliminating the sun flare, in fact, water glitter sometimes has useful information of water body. This information may play an important role for the water remote sensing retrieval. For example, the mirror reflection produced by the oil slick on water surface, and this information will be lost if we completely eliminate the glitter. In practice, the water body scattering is very weak, after completely eliminating the glitter, the water information received by the sensors will be too little to be detected because of the polarizer's absorption. On the other hand, it is hard to let the sensors' detection angle state at the Brewster angle. So, it is impossible and no need to completely eliminate the sun flare. As long as the sensor has not saturated by the glitter's radiation, the upward radiation of water will be identified effectively. That is to say, when the incident angle is nearby the Brewster angle, we can adjust the glitter's radiation by controlling the polarizer's azimuth. In this way, we can eliminate the noise and enhance the useful information.

Using different sensors, we can quantitatively present the suitable solar incident angle and the suitable polarization azimuth for the polarization remote sensing. It can keep the intensity of sun flare in an acceptable range and give the suitable schedule for the water polarization remote sensing. The suitable times will be much longer than the case that completely eliminates the sun flare. The sensor will receive more water information [17].

2.5. The polarized effect on plant spectrum

Leaf scattering spectrum is one of the key optical variables that conveys information about leaf absorbing constituents from remote sensing. It cannot be directly measured from space because the radiation scattered from leaves is affected by the three-dimensional canopy structure. In addition, some radiation is specularly reflected by the surface of leaves. This portion of reflected radiation is called partly polarized. It does not interact with pigments inside the leaf and therefore contains no information about its interior. Very few empirical data are available on the spectral and angular scattering properties of leaf surfaces. Whereas canopy structure effects are well understood, the impact of the leaf surface reflectance on estimation of leaf absorption spectra remains uncertain. We, thus, present empirical and theoretical analyses of spectral, angular, and polarimetric measurements of light reflected by needles and shoots of *Picea koraiensis* and *Pinus koraiensis* species. **Figure 6** illustrates our samples.

Figure 6. Samples of needles and shoots in the holder window. Sizes of the shoots were 13 × 15 × 17 cm (*Pinus koraiensis*) and 6 cm (*Picea koraiensis*). Dimensions of the holder windows were 5.3 × 7 × 14 cm (*Pinus koraiensis*) and 5.0 cm (*Picea koraiensis*).

The total radiation reflected by a leaf includes two components, diffuse and specular. The first component emitting from light reflected at the air-cuticle interface is polarized. The diffuse component results from photon interactions within the leaf and large particle on the leaf surface. This portion of reflected light is not polarized. Polarization measurements can help us to extract linearly polarization portion from the total radiation registered by the sensor. Radiation specularly reflected from the needle surfaces exhibits weak spectral dependency, as expected from theory. It increases from very small values in backscattering directions to about 17% in forward scattering directions. The shoot sample, polarized directional-conical reflectance factor (PDCRF), shows a similar phenomenon. Its magnitude, however, is reduced by a factor of about 10, as **Figure 7** shows.

Ignoring polarization portion in reflected radiation, however, can cause an overestimation of the scattering coefficient (**Figure 8**). The impact decreases from strongly (17–140%, 450–500 nm) to weakly (<4%, 800–950 nm) absorbing wavelengths.

To summarize, the spectral, angular, and polarimetric data convey information about properties of the needle surfaces, shoot structural organizations, and needle optics. This information is required to retrieve the needle albedo, which is directly related to the absorption spectra of leaf biochemical constituents [18].

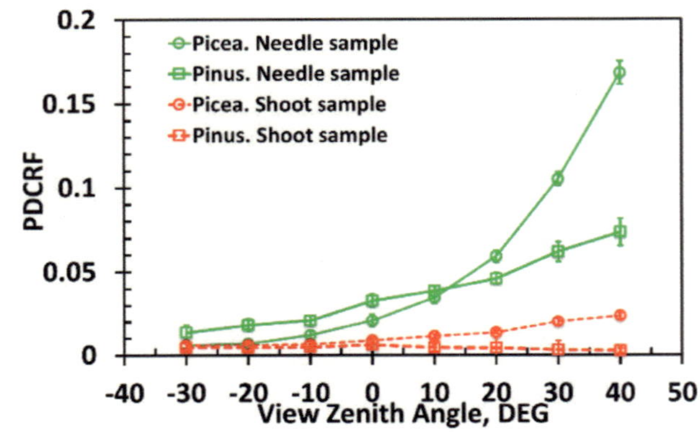

Figure 7. Angular distribution of average PDCRF of shoot samples (dashed lines) and needle samples (solid lines) averaged over 450–950 nm. Vertical bars denote ±1 standard deviation.

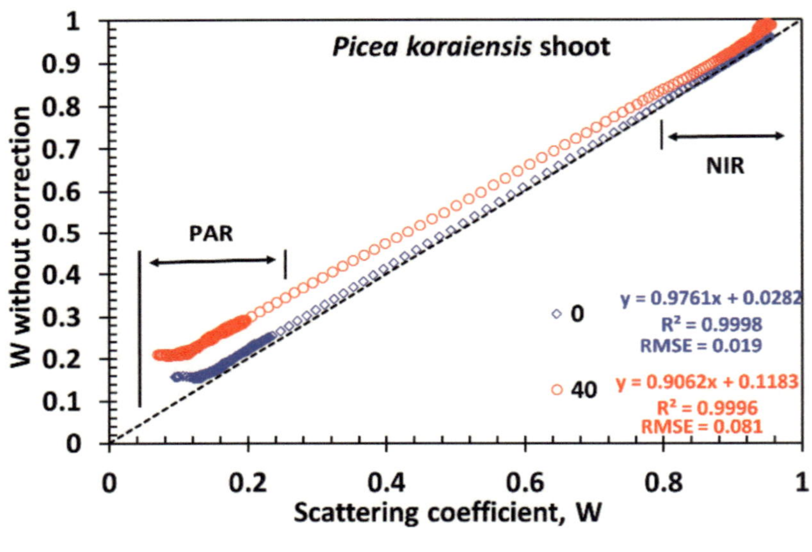

Figure 8. Correlation between scattering coefficients of the *Picea koraiensis* shoot derived with (horizontal axis) and without (vertical axis) correction for the needle surface effects. Relative differences are 17–140% in 450–500 nm, 3–74% in 600–650 nm, 3–59% in 520–580 nm, and below 4% in 800–950 nm spectral intervals.

3. Polarized remote sensing for atmospheric correction

Scattered atmospheric particles exhibit strong polarization phenomena. The polarization effect of the atmosphere is the main signal of the polarization remote sensing. For this reason, the present spaceborne polarization remote sensing data are mainly used for atmospheric research, such as to study the atmosphere physical properties and optical properties [19]. Land objects also have strong reflected polarization phenomenon and can be valuable information for remote sensing.

The land objects' detection was one of the tasks of the spaceborne polarimeter—POLarization and Directionality of the Earth's Reflectance (POLDER). However, studies found that the polarization effect of the atmosphere is much stronger than that of land objects in the images. Information on land object polarization received by the polarimeter is always submerged in the atmospheric polarization effect [20, 21]. As a result, a major concern for using polarized light for the study of land surfaces is the capability to discriminate between polarization generated in the atmosphere and that generated by the surface.

The neutral point is the point (or area) where the skylight is unpolarized. In the clear sky, there are three normally occurring neutral points, the Arago, Babinet, and Brewster neutral points, in the principal plane [22]. In this paper, we attempt to set spaceborne polarimeter to detect the Earth at the direction of the neutral point. Because the polarization effect of atmosphere is zero at this direction, the polarization information of the land surface objects can be maximized. This study would promote the polarized remote sensing for land objects detection and expand the polarization remote sensing research to a wider research area.

3.1. Atmospheric polarization distribution and neutral point

The solar radiation has no polarization at the outer space, and it will be polarized after the atmosphere particles is scattered. If most of the atmosphere scattering is single scattering, the polarization of the sky will show a regular polarization distribution, and it is also known as the polarization pattern of sky.

Under the clear sky weather conditions, in the vertical plane of sun, the Arago, Babinet, and Brewster neutral points in the sky will appear. As shown in **Figure 9**, the Arago is normally

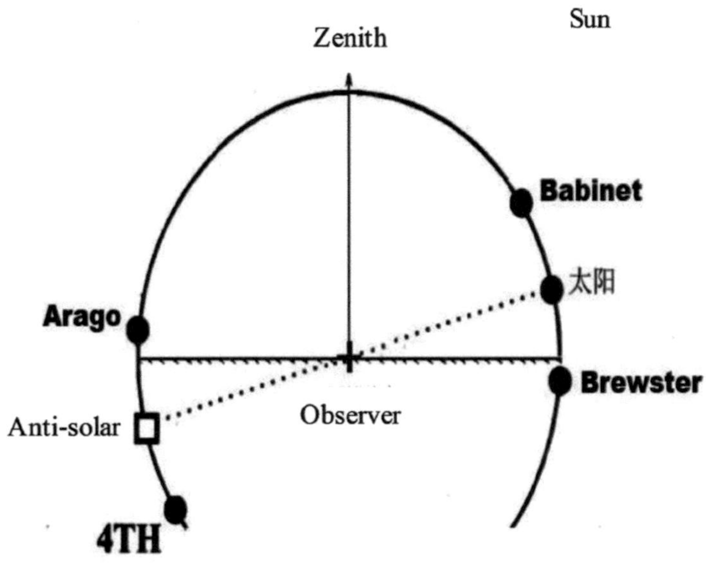

Figure 9. Relative positions of neutral points in the sky.

located 20–30° above the antisolar point. The Babinet point is directly above the sun located at 15–20°. The Brewster point is directly below the sun varying from 15–20°. Both the Babinet and Brewster points are at the same direction of sun and their positions change along with the solar altitude. The higher the sun elevation angles, the closer between the Babinet point and the Brewster point. When the sun is at the zenith, the two points merged as one point [23, 24]. Because the neutral points at two side of sun and antisolar are symmetrical, there should be another neutral point under antisolar in theory, as the anti-Babinet neutral point. In 2002, the fourth neutral point actually has been detected using the sounding balloon [25].

For the single scattering, the polarization of the skylight is generally positive. However, for the multiple scattering, the atmospheric particles can cause negative polarization. The degree of polarization of sky will be zero where the positive and negative polarizations meet at the intersection area. In this way, the atmospheric neutral point is produced [26]. The neutral point is always at the main plane of the sun and zenith. The stronger the multiple scattering, there will be more negative polarization, and the neutral point position will be farther away from the theory position.

The positions of neutral points are close to the sun elevations. Chandrasekhar calculated positions of the three neutral points for various angles of incidence in the main plane. **Figure 10** shows the case for atmospheric optical thickness of 0.10. The abscissa and ordinate give the solar elevation angle and neutral point elevation angle, respectively.

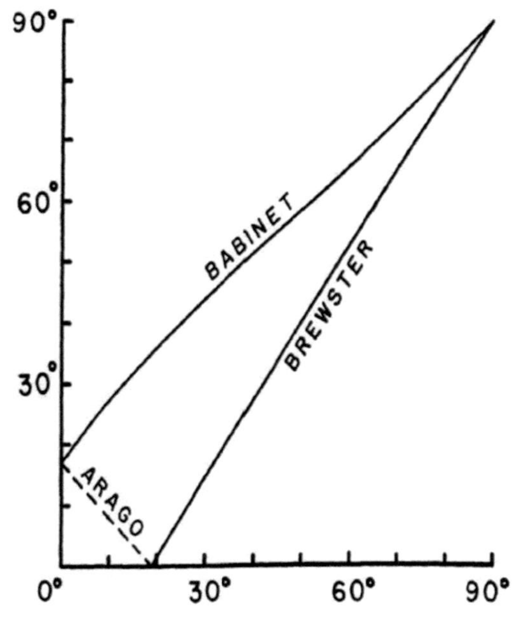

Figure 10. The relationship of neutral point and sun position.

Babinet point is visible throughout the day from before sunrise until after sunset. The Brewster neutral point becomes visible when solar elevation angle exceeds 20°. The position of the Brewster point and sun is almost linear. As for Arago point, it will appear when solar elevation angle is less than 20°. It means that the Arago point can only be observed in the morning or evening. The situation of other atmospheric optical thicknesses (0.15 or 0.20) is almost the same.

3.2. Atmosphere correction method based on neutral point

The method of using atmospheric neutral point for polarization remote sensing is to set the remote sensing sensor at the direction of the neutral point. At this moment, the atmospheric polarization effect between sensor and land objects can be zero or minimized. The polarization information of land objects received by sensor can be maximized. **Figure 11** is the sketch map of using neutral points for polarization remote sensing atmosphere correction. The dash circle is virtual outside the Earth's atmosphere. This circle is also the sun trail of the very day. The sun position in the figure is a virtual position too, which only denotes the direction of the sun in the figure. From space to land surface, the neutral point can be seen at each height in the line that goes through the neutral point and land object in theory. So, the neutral point can be observed through by airborne or spaceborne sensors.

In the real atmosphere, because of multiple scattering, the neutral point is probably not a point but a small region where all the degree of polarization is close to zero. In addition, this neutral point (region) in the sky is not a fixed position, but a conical region where starting point is the sensor as shown in **Figure 11**. The atmosphere degree of polarization at any height of this conical region is zero.

3.2.1. Neutral point characteristics observed from space

The positions of neutral points that we discussed above are based on ground observations; however, remote sensing is a process observed from sky to land. Can the remote sensing sensor

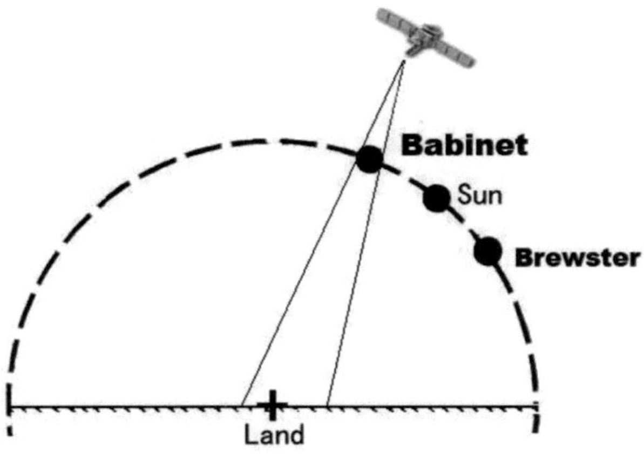

Figure 11. Sketch map of using neutral points for polarization remote sensing atmosphere correction.

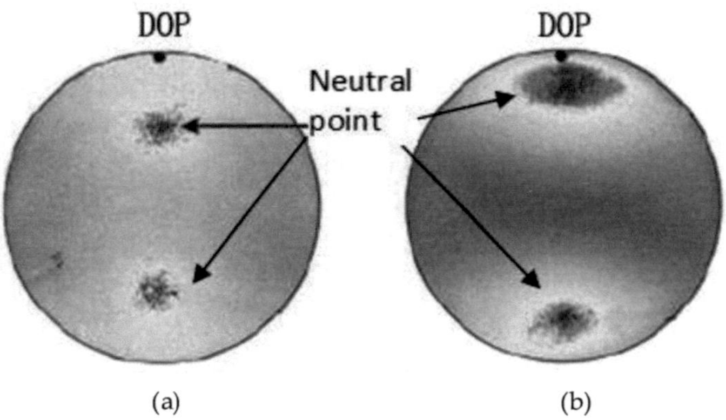

Figure 12. The neutral point observed from ground and space (the black areas are neutral points where DOP is nearly zero). (a) observed from ground, and (b) observed from space.

detect the neutral point from sky? It is the basic problem of using the neutral point for land surface polarization remote sensing.

Gábor Horváth [27] took the neutral point photos on the ground and at a 3500 m sounding balloon separately, as the dark areas shown in **Figure 12**. The local solar elevation angle shooting was similar to 0 when photos were taken. The black spot at the top of the image is the sun. The camera band was red. **Figure 12(a)** was the ground photo, while **Figure 12(b)** was the balloon photo. The right picture of **Figure 12(b)** indicates that the neutral point can be observed from space.

Coulson calculated the neutral point position of the atmosphere upward radiation and the downward radiation [28]. **Figure 13** shows the curve of angular distance between the sun and the neutral points as a function of sun elevation. The abscissa is the solar elevation angle and the ordinate is angular distance from neutral to antisolar. The atmosphere optical thickness is 1.0. The solid curves are the upward radiation and the dash curves are the downward radiation.

The upward radiation is caused by the backscattering of atmospheric particles. The neutral point's positions are different in the upward radiation and the downward radiation. So, the neutral points have different names in the upward radiation and the downward radiation. The neutral points observed from ground are Arago point, Babinet point, and Brewster point. And the corresponding neutral points observed from the space are Brewster point, anti-Babinet point, and Arago point.

In **Figure 13**, the solid and dash lines of curve (a) denote the neutral point below the antisolar and solar separately, which are Arago point and Brewster point. The solid and dash lines of curve (b) denote anti-Babinet point and Babinet point separately.

It can be seen from **Figure 13**, as for curve (a), that the neutral point positions in the upward and downward radiation are completely the same when the solar elevation angle is about 60°–90°. Their positions began to differ greatly when the solar elevation angle changed from

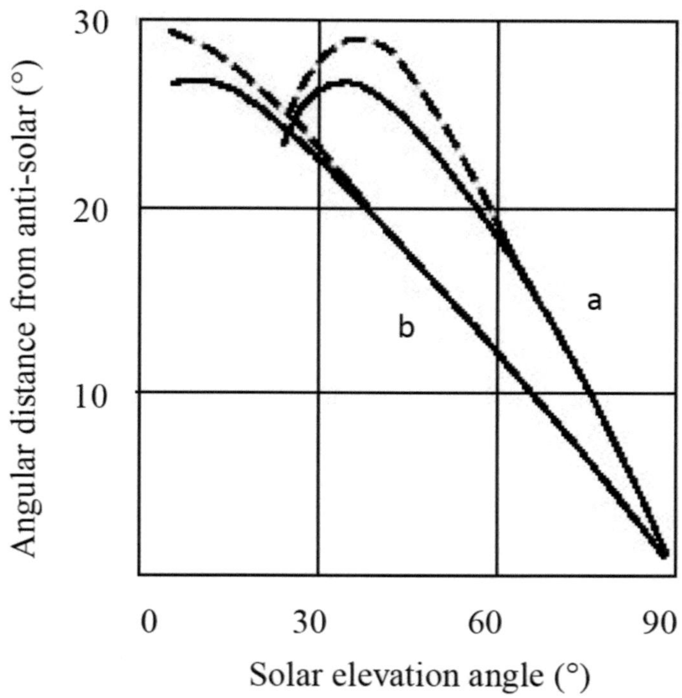

Figure 13. Angular distance between the sun and the neutral points.

25 to 60°. Both the neutral points in the upward (Arago point) and downward (Brewster point) radiation cannot appear in the sky when the solar elevation angle is less than 25°. As for curve (b), the two neutral points' positions in the upward (anti-Babinet point) and downward radiation (Babinet point) are entirely consistent when the solar elevation angle is about 32°–90°. Only when the solar elevation angle is small (about 5°–32°), the positions began to be different slightly.

The above discussion shows that the neutral point positions of the same area observed from space and ground are consistent when the atmosphere optical thickness is small and the solar elevation angle is big. Normally, satellite transit time or aerial remote sensing flight time always choose a high solar elevation, in order to obtain sufficient light conditions. At this time, the space-based neutral point position can be calculated by the ground-based observation position.

Kattawar compared the neutral point positions of different atmosphere optical thicknesses in the upward and downward radiations [29]. The result shows that when the atmosphere optical thickness increased, the Babinet point position changed slightly while the Brewster point changed greatly. The position between anti-Babinet point in the upward radiation and Babinet point in the downward radiation is almost the same when the atmosphere optical thickness is <5. This also indicates that the Babinet point positions of the ground-based and space-based observations are coincident of the same area. This paper also gives the impact of the surface albedo on the neutral point position. The change of surface albedo has small effect to the neutral point position. But the Brewster point is more sensitive than the Babinet point on the surface albedo impact. The anti-Babinet point in the upward radiation is independent of surface albedo.

3.2.2. Proper neutral point selected for remote sensing

There are three common neutral points in the sky. We need to select the most applicable one for the polarization remote sensing. The first satisfied condition is its position in the sky, convenient for remote sensing observation. The second condition is that its position should be more stable.

In the visible-near-infrared remote sensing, satellite transits time or aerial remote sensing flight time always chooses a higher solar elevation, in order to obtain sufficient light conditions. The Arago neutral point can only be observed in the morning or toward evening, and its position in the sky is low when appearing. Its position is always lower than sun. Especially, in high latitude area, the sun cannot reach the zenith position. The Arago point is not suitable for polarized remote sensing. The Brewster neutral point has the same situation. The Brewster point position is bound to be even lower. In addition, the Brewster point is affected by the atmosphere condition and land surface reflection greatly. So, the Brewster neutral point is also not suitable for polarization remote sensing.

In contrast, the anti-Babinet neutral point is an ideal choice for polarization remote sensing. From the earth observing characters of Babinet neutral point, the anti-Babinet point is in the same side of the sun; its position is always higher than the sun. It means that the anti-Babinet point can be observed all the day. Higher elevating angle is a benefit in remote sensing to obtain better light conditions. Moreover, the anti-Babinet point is not sensitive to the atmosphere condition. Its position is more stable in the sky when the atmosphere condition changes.

3.2.3. The position of neutral point calculated

The position of the anti-Babinet neutral point can be calculated from the position of the Babinet neutral point, since the two points almost have the same position.

The main factor of the anti-Babinet point position is the solar elevation. So when using anti-Babinet point for land surface polarization observation, the first thing is to determine the observation time.

The solar elevation angle is constantly changing at the same place a day. The sun elevation angle of a place can be calculated by the solar elevation angle formula, as in formula (13).

Generally, good weather conditions are selected for aerial remote sensing flight. For the anti-Babinet neutral point, its position in the main plane is the function of the sun elevation when the optical thickness is 0.10. For convenience, the curve was separated into two parts. Both curves fit as segments.

$$\begin{cases} y = 0.9x + 18 & (0 < x < 30) \\ y = 0.75x + 22.5 & (30 <= x < 90) \end{cases} \tag{14}$$

where x is the solar elevation angle and y is the Babinet neutral point elevation angle.

For the equatorial region, the solar elevation angle of 90° at noon, the sun illuminates the land surface vertically. All the neutral points are gathered into the sun position. The polarized remote sensing sensor should be in the neutral point region when observing land surface vertically, and this is a special case.

For nonequatorial regions, the sun elevation angle cannot reach 90° at noon. In local noon, the solar elevation angle can be calculated by formula (13). Because both neutral point and sun are at the main plane, the corresponding neutral point position can be calculated from formula (14), and then we can get the remote sensing time. One thing needs to be noted that even the sensor in the sky cannot be exactly at the center of the neutral point, and the region around the neutral point is also a small polarization area; it is also conducive to polarization observation.

3.2.4. Experimental verification

In order to verify the method of using the neutral point for polarization remote sensing, we designed a ground verification experiment. Polarization images were taken and compared from neutral point direction and non-neutral point direction for the same area. **Figure 14** is the observation geometry sketch map. **Figure 14(a)** is the sketch map of neutral point direction observation, while **Figure 14(b)** is the sketch map of non-neutral point direction observation. The degree of polarization of atmosphere is zero on the line through the neutral point and camera. The imaging device was Nikon D200 digital camera with iodine polarizer. 0, 60, and 120°, three angle polarization images, were obtained by changing the angle between polarizer transmission axis and reference axis three times. And then the images of polarization parameters such as degree of polarization and polarization angle were calculated.

Figure 14 is the polarization parameter comparison images between non-neutral point imaging and neutral point imaging. The observation time was 7:20 am and 11:20 am separately, April 29, 2010. At that time, the solar elevation angle was 24.40 and 62.95° separately. According to formula

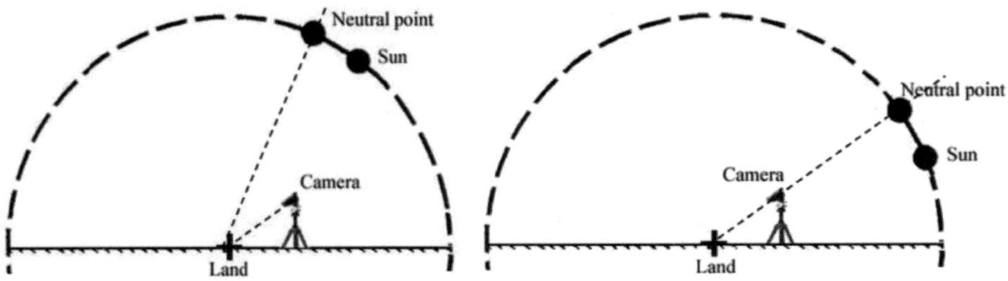

Figure 14. Observation geometry sketch map of ground verification experiment. ◎a◎Observation at non-neutral point direction (in the noon) (b) Observation at neutral point direction (in the morning).

(14), the neutral point elevation angle was 69.71 and 39.96° separately. The solar azimuth was 21.70 and 88.55° (the south is defined as the zero azimuth) separately. Both the camera detection elevation angle was set as 40°. The relative azimuth angle of the camera and the sun were 0° (backward) and 66.85°, respectively. The weather was clear and calm. Observation spots were fifth floor platform, remote sensing building, and Peking University. The target area is Summer Palace scenic region. **Figure 14(a)** is the non-neutral point direction observation. **Figure 14(b)** is the neutral point direction observation.

Figure 15(a) shows the images without polarizer. The atmosphere visibility is high. Both the images are clear: whether observation at the neutral point direction or non-neutral point direction. **Figure 15(b)** shows the linear degree of polarization (*DoLP*) images of R, G, B bands. Degree of polarization is a physical quantity that reflects the polarization size of objects. It is a dimensionless number and the value is 0–1. The white and black areas in the figure denote the big value of *DoLP* and small value of *DoLP* separately. On the image taken from neutral point direction (right image), whether the closer tower of Buddhist Incenses (about 3.1 km from camera) or the Forane Mountain (about 8 km from camera) are clearly visible, especially the bare soil and road in the mountains. And the polarization information is not the same in the red, green, and blue bands. However, on the image taken from non-neutral point direction (left image), the closer objects are better displayed, such as the tower of Buddhist Incense. But the polarization information of the forane objects is much weaker, only the mountain outline can be shown. It means that the polarization information of forane objects cannot be obtained from the non-neutral point direction image. This phenomenon shows that along with the increase of the focusing distance, the atmospheric polarization effects increases, and the polarization information of targets becomes weak.

The information entropies were calculated for the unpolarized images in **Figure 15(a)** and the three bands of *DoLP* images in **Figure 15(b)**, as shown in **Figure 16(a)**. The abscissa is different images and the ordinate is the value of information entropy. The twill denotes entropy value of the non-neutral point imaging and the plaid denotes entropy value of the neutral point imaging. It can be seen from **Figure 16(a)** that the information entropies of the two unpolarized images are almost equal regardless of the neutral point imaging or non-neutral point imaging. However, for the *DoLP* images of red, green and blue bands, the information entropy values of the neutral point imaging are all greater than the non-neutral point imaging. This also shows that land objects on the neutral point *DoLP* imaging are informative. In particular, we cut out the upper half of all the images, keep only the distant objects on the images, and then the information entropies were calculated separately, and the results are showed in **Figure 16(b)**. Similarly, the information entropies of the two nonpolarized images are almost the same. But for the *DoLP* images of red, green, and blue bands, the information entropy values of the neutral point imaging are all much greater than the non-neutral point imaging.

It is thus clear that the atmospheric polarization effect on the non-neutral point direction image is stronger than the neutral point direction image. The degree of polarization image from neutral point direction has good expressive force for remote objects. The objects' polarization information on it is far greater than on non-neutral point direction image's. It means the neutral point imaging can eliminate the atmospheric polarization effect and enhance the objects' polarized information. And then, we had a polarized remote sensing aerial flight experiment with atmospheric neutral

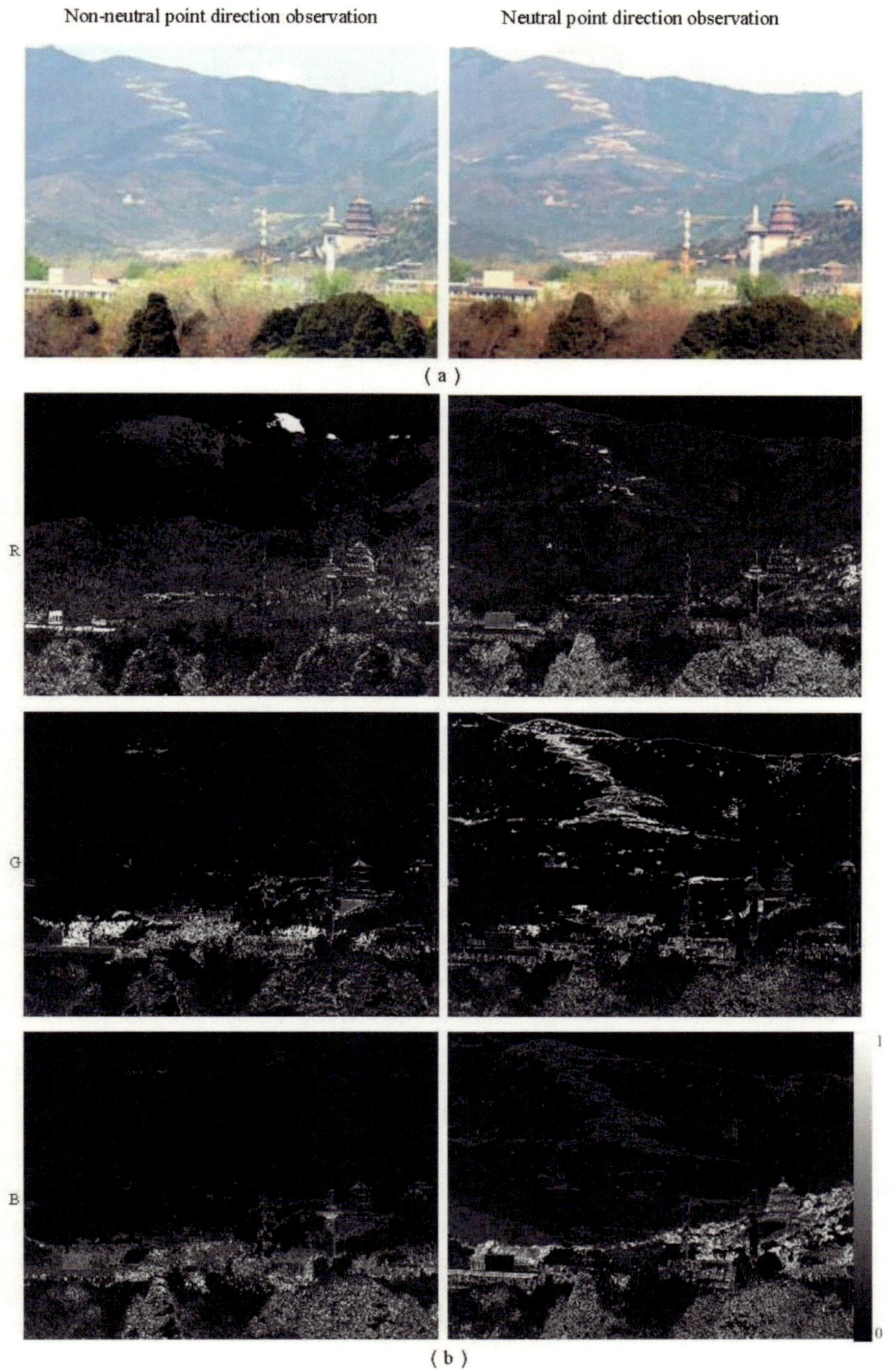

Figure 15. The polarization collation images between neutral point imaging and non-neutral point imaging. (a) Images taken without polarizer and (b) Degree of polarization images.

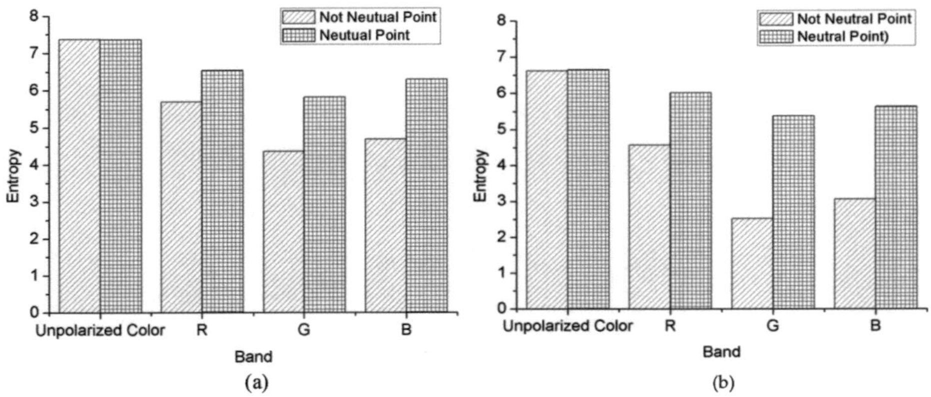

Figure 16. The information entropy comparison between non-neutral point and neutral point imaging.

point [30]. This experiment demonstrated the feasibility of using neutral point for polarized remote sensing in atmospheric correction [31].

4. Conclusion

In conclusion, this chapter showed the researches on land object polarization properties, which can provide a new base knowledge for polarization remote sensing and a new research breakthrough in the separation method for the polarization effect between objects and atmosphere.

1. The polarized bidirectional reflectance characteristics and polarized hyperspectral properties of land objects were methodically studied. It is attempted to find the object polarization reflectance mechanisms by the measurements and theoretical derivations. The results showed that the land object polarization reflectance had the law of bidirectional reflectance and that there was a quantitative relationship between the bidirectional reflectance and the polarized bidirectional reflectance. These two rules can provide the theoretical basis for polarization remote sensing such as the detecting conditions, modeling, and others. This chapter also gave the polarized spectral property of the typical objects. It can be also used as the spectral basis for polarization remote sensing.

2. It proposed a method of using atmospheric neutral point for the separation polarization effect between objects and atmosphere. In this study, we attempted to install the polarization sensor at the direction of the atmosphere point, at the neutral point. In this case, the polarization effect of the atmosphere was reduced to zero and the polarization information of the land surface was maximized. The theoretical derivation and ground experimental results indicated the feasibility of using atmosphere neutral point to separate the polarization effect between object and atmosphere.

Author details

Lei Yan[1], Taixia Wu[2*] and Xueqi Wang[1]

*Address all correspondence to: wutx@hhu.edu.cn

1 School of Earth and Space Sciences, Peking University, Beijing, China

2 School of Earth Sciences and Engineering, Hohai University, Nanjing, China

References

[1] Horváth G, Gál J, Pomozi I, Wehner R. Polarization portrait of the Arago point: Video-polarimetric imaging of the neutral points of skylight polarization. Naturwissenschaften. 1998;**85**(7):333-339. DOI: 10.1007/s001140050510

[2] Wang Y, Yang C, Li X. Regularizing kernel-based BRDF model inversion method for ill-posed land surface parameter retrieval using smoothness constraint. Journal of Geophysical Research-Atmospheres. 2008;**113**(D13):1-11. DOI: 10.1029/2007jd009324

[3] Taixia W, Yunsheng Z. The bidirectional polarized reflectance model of soil. IEEE Transactions on Geoscience and Remote Sensing. 2005;**43**(12):2854-2859. DOI: 10.1109/tgrs.2005.857905

[4] Herman M, Deuzé JL, Devaux C, Goloub P, Bréon FM, Tanré D. Remote sensing of aerosols over land surfaces including polarization measurements and application to POLDER measurements. Journal of Geophysical Research-Atmospheres. 1997;**102**(D14):17039-17049. DOI: 10.1029/96JD02109

[5] Kochenova S, Lerot C, De Maziere M, Letocart V. Importance of polarization for remote sensing of the Earth's atmosphere. In: EGU General Assembly Conference. 2012

[6] Deschamps P-Y, Bréon F-M, Leroy M, Podaire A, Bricaud A, Buriez J-C, et al. The POLDER mission: Instrument characteristics and scientific objectives. IEEE Transactions on Geoscience and Remote Sensing. 1994;**32**(3):598-615. DOI: 10.1109/36.297978

[7] Talmage D, Curran P. Remote sensing using partially polarized light. International Journal of Remote Sensing. 1986;**7**(1):47-64. DOI: 10.1080/01431168608954660

[8] Shibayama M, Watanabe Y. Estimating the mean leaf inclination angle of wheat canopies using reflected polarized light. Plant Production Science. 2007;**10**(3):329-342. DOI: 10.1626/pps.10.329

[9] Schott JR. Fundamentals of Polarimetric Remote Sensing. Vol. 81. Spie Press; 2009

[10] Yun-sheng Z, Tai-xia W, Bao L, Yang-jie L. Quantitative relationship between multi-angle polarized reflectance and BRDF of rock. Journal of China University of Mining and Technology. 2005;**15**(3):192-196

[11] Cloude S. Polarisation: Applications in Remote Sensing. Oxford University Press; 2010

[12] Zhao Y, Wu T, Hu X, Luo Y. Study on quantitative relation between multi-angle polarized reflectance and bidirectional reflectance. Journal of Infrared and Millimeter Waves-Chinese Edition. 2005;**24**(6):441. DOI: 10.3321/j.issn:1001-9014.2005.06.010

[13] Tonizzo A, Zhou J, Gilerson A, Twardowski MS, Gray DJ, Arnone RA, et al. Polarized light in coastal waters: Hyperspectral and multiangular analysis. Optics Express. 2009; **17**(7):5666-5683. DOI: 10.1364/OE.17.005666

[14] Aron Y. Polarization in the LWIR: A method to improve target aquisition. In: Proceedings of SPIE 5783, Infrared Technology and Applications XXXI. Vol. 5783. 2005. pp. 653-661. DOI: 10.1117/12.605316

[15] Schau HC. Estimation of polarization state from hyperspectral imagery. In: Proceedings of SPIE 5425, Algorithms and Technologies for Multispectral, Hyperspectral, and Ultraspectral Imagery X. Vol. 5425. 2004. pp. 470-479. DOI: 10.1117/12.541719

[16] Wu T, Zhang L, Cen Y, Huang C, Sun X, Zhao H, et al. Polarized spectral measurement and analysis of Sedum spectabile Boreau using a field imaging spectrometer system. IEEE Journal of Selected Topics in Applied Earth Observations and Remote Sensing. 2013;**6**(2): 724-730. DOI: 10.1109/JSTARS.2013.2255862

[17] YangJie L, YunSheng Z, XiaoWen L, TaiXia W, LiLi Z. Research and application of multi-angle polarization characteristics of water body mirror reflection. Science in China. Series D, Earth Sciences. 2007;**50**(6):946-952. DOI: 10.1007/s11430-007-0009-9

[18] Yang B, Knyazikhin Y, Lin Y, Yan K, Chen C, Park T, et al. Analyses of impact of needle surface properties on estimation of needle absorption spectrum: Case study with coniferous needle and shoot samples. Remote Sensing. 2016;**8**(7):563. DOI: 10.3390/rs8070563

[19] Fan X, Goloub P, Deuzé JL, Chen H, Zhang W, Tanré D, et al. Evaluation of PARASOL aerosol retrieval over North East Asia. Remote Sensing of Environment. 2008;**112**(3):697-707. DOI: 10.1016/j.rse.2007.06.010

[20] Bicheron P, Leroy M, Hautecoeur O, Bréon FM. Enhanced discrimination of boreal forest covers with directional reflectances from the airborne polarization and directionality of earth reflectances (POLDER) instrument. Journal of Geophysical Research-Atmospheres. 1997;**102**(D24):29517-29528. DOI: 10.1029/97JD01330

[21] Nadal F, Breon FM. Parameterization of surface polarized reflectance derived from POLDER spaceborne measurements. IEEE Transactions on Geoscience and Remote Sensing. 1999; **37**(3):1709-1718. DOI: 10.1109/36.763292

[22] Miyazaki D, Ammar M, Kawakami R, Ikeuchi K. Estimating sunlight polarization using a fish-eye lens. Information and Media Technologies. 2010;**5**(1):164-176. DOI: 10.11185/imt.5.164

[23] Sheng P, Mao J, Li J. Atmospheric Physics. Beijing: Peking University Press; 2003

[24] Muheim R, Phillips JB, Åkesson S. Polarized light cues underlie compass calibration in migratory songbirds. Science. 2006;**313**(5788):837-839. DOI: 10.1126/science.1129709

[25] Barta A, Bernáth B, Suhai B, Horváth G, Wehner R. First observation of the fourth neutral polarization point in the atmosphere. Journal of the Optical Society of America. A, Optics, Image Science, and Vision. 2002;**19**(10):2085-2099. DOI: 10.1364/JOSAA.19.002085

[26] Horváth G, Varjú D. Polarized Light in Animal Vision: Polarization Patterns in Nature. Springer Verlag; 2004

[27] Horváth G, Bernáth B, Suhai B, et al. First observation of the fourth neutral polarization point in the atmosphere.[J]. Journal of the Optical Society of America A Optics Image Science & Vision. 2002;**19**(10):2085-2099

[28] Coulson K. Characteristics of the radiation emerging from the top of a rayleigh atmosphere—I: Intensity and polarization. Planetary and Space Science. 1959;**1**(4):265-276. DOI: 10.1016/0032-0633(59)90032-7

[29] Kattawar GW, Plass GN, Hitzfelder SJ. Multiple scattered radiation emerging from Rayleigh and continental haze layers. 1: Radiance, polarization, and neutral points. Applied Optics. 1976;**15**(3):632-647. DOI: 10.1364/AO.15.000632

[30] Yang S, Guan G, Zhao H, Zhao H, Yang B, Zhang W, et al. Airborne validation of ground-object detection from polarized neutral-point atmosphere. Spectroscopy and Spectral Analysis. 2013;**33**(9):2525-2531. DOI: 10.3964/j.issn.1000-0593(2013)09-2525-07

[31] Wu T, Zhang L, Cen Y, Huang C, Zhao H, Sun X. Neutral point consideration for atmospheric correction in polarization remote sensing. Journal of Remote Sensing. 2013;**17**:235-247. DOI: 10.11834/jrs.20132156

Section 3

Numerical Weather Prediction

Chapter 4

Rainfall Nowcasting by Blending of Radar Data and Numerical Weather Prediction

Hai Chu, Mengjuan Liu, Min Sun and Lei Chen

Additional information is available at the end of the chapter

http://dx.doi.org/10.5772/intechopen.81632

Abstract

In order to improve conventional rainfall nowcasting, radar extrapolation and high-resolution numerical weather prediction (NWP) were blended to get a 6-h quantitative precipitation forecast (QPF) over the Yangtze River Delta region of China. Modifications and calibrations were done to both the extrapolation and NWP in order to get an integrated result from the two, which mainly included the extension for the extrapolation time and region, intensity and position calibration for the NWP, weighted blending of extrapolation and NWP based on scale and time, and a final real-time Z-R relation conversion. Forecast experiments were done, and results show that the blending technique could effectively extend forecast time compared with conventional radar extrapolation, meanwhile applying a positive calibration to the NWP. The overall CSI score of 0–6 h reflectivity forecast was better than either single forecast.

Keywords: nowcasting, radar extrapolation, NWP calibration, blending forecast

1. Introduction

Practice has shown that deterministic short-term extrapolation forecasts have its predictive capability for forecasting precipitation within a few hours. Using modern high-resolution observation, combined with simple extrapolation model, a satisfactory precipitation prediction result with high time-space resolution can be obtained. When forecast time increases to 2–6 h, optimal prediction result can be obtained by combining the results of short-term extrapolations and numerical models together, and the blended product will be superior to any of the two [1–3].

As to various kinds of extrapolation methods, the main difference often lies in the acquisition of extrapolation vectors, and there are mainly three of them: The first one is to determine the position of the same echo at different times by using correlation method. The second one is to use echo recognition tracking technology. By decomposing the precipitation echo into different independent echo units, moving vectors of each unit can be obtained by matching units of adjacent times. The third one is to use observation wind field or NWP forecast wind field directly to carry on the echo extrapolation. All the three methods have shown their advantages in operation, such as the models of TITAN [4], TREC [5, 6], and optical flow [7, 8]. At present, both TITAN and TREC are in operation here at Shanghai Central Meteorological Observatory (SCMO), and of which the modified COTREC method is more commonly used by forecasters [9].

As to forecasts of high-resolution regional numerical models, when they are applied directly to short-term prediction, several aspects need to be considered: The first one is the computation time, as NWP model takes time to do data assimilation and parallel integral computation; the result of the first 1–2 h of integral is often not available in practice. When time increases, its forecast will inevitably deviate from the current reality, and its results will need to be corrected. The second one is the resolution, mainly spatial resolution. The common resolution of regional numerical model is about 3–10 km, which is enough to reflect the small and mesoscale weather system, but it is still rough compared to the radar extrapolation of 1 km, particularly in the case of complex urban boundaries; high-impact disaster weather tends to occur within a few dozen square kilometers, and NWP is generally considered not able to effectively distinguish a system with a size less than five times of its resolution. Therefore, it is necessary to apply calibration to conventional NWP products.

Finally, as to the short-term quantitative precipitation forecast, the prevailing view is that to use the combination of conventional short-term forecasting techniques (mainly radar extrapolation) and numerical forecasts is a direct way to increase forecast to more than 2 h. The Nimrod system in the UK, for example, makes short-term strong precipitation forecasts by giving extrapolation and NWP different weights according to forecast time [10]. The NIWOT system from NCAR, on the other hand, tries to modify the size of extrapolation using information from NWP [11]. In the Swirl System of the Hong Kong Observatory (HKO), the prediction of NWP is revised by means of phase correction and strength correction, and a hyperbolic function is used to determine the weight of extrapolation and NWP, thus effectively improving the prediction effect of 0–6 h [12]. Due to the relatively mature swirl system and good openness, HKO has now collaborated with several advanced local bureaus in China such as the Beijing Meteorological Service and the Zhejiang Meteorological Bureau on the task of blending nowcasting [13, 14].

In this study, the authors also try to explore a set of short-term forecasting blending methods for the Yangtze River Delta region by using advanced radar network observation in East China and combining high-resolution numerical model forecast. In the following part of this chapter, we first briefly introduce the extrapolation method, and then we will focus on how to set the system configuration and calibrate the final forecast of NWP to obtain a better blending with the extrapolation. Conclusion and discussion are given in the end.

2. Blending scheme

The aim of this study is to establish a blended quantitative precipitation product for 0–6 h short-term forecast, covering the Yangtze River Delta region with spatial resolution of 3 km and time resolution of 10 min. Using high-resolution radar network observation extrapolation and specially configured numerical model, blended product is then carried out by real-time model evaluation and calibration, so as to improve the timeliness and accuracy of prediction (**Figure 1**).

2.1. Extended 0–6 h radar extrapolation

In this chapter, an improved TREC method, namely, the COTREC [9], is used for the extrapolation. First, a level-II real-time data quality control is done to remove non-meteorological echoes such as super refraction and geophysical echo. Non-meteorological echoes are mostly concentrated in low elevation angle; thus, it can be eliminated by doing comparative analysis of different elevation data, which in this study is 0.5°/1.5° VCP scan. Then, cross-correlation method (TREC) is used to find the TREC wind field. The wind field is first smoothed out by adopting a 9-point smoothing to avoid sudden change. As wind field retrieved by TREC method is only of value in the region where there is observed reflectivity, so the non-convergence wind field obtained by TREC is usually weakened to a certain extent, which is particularly the case when it comes to isolated echoes or lined echoes. By introducing wind field from numerical forecast as the guide flow field of TREC wind field, the weakening of TREC wind field can be compensated. In this chapter, a 3-hourly forecast wind field of GFS, a global forecast model provided by the National Centers for Environmental Prediction (NCEP) of the USA, is used to supply the TREC wind field. The missing values of TREC wind is compensated by doing vertical averaging and horizontal interpolation of the GFS wind. Thus, a new TREC wind field (nu0 (i, j), nv0 (i, j)) is obtained. The final wind field of COTREC (u (i, j), v (i, j))

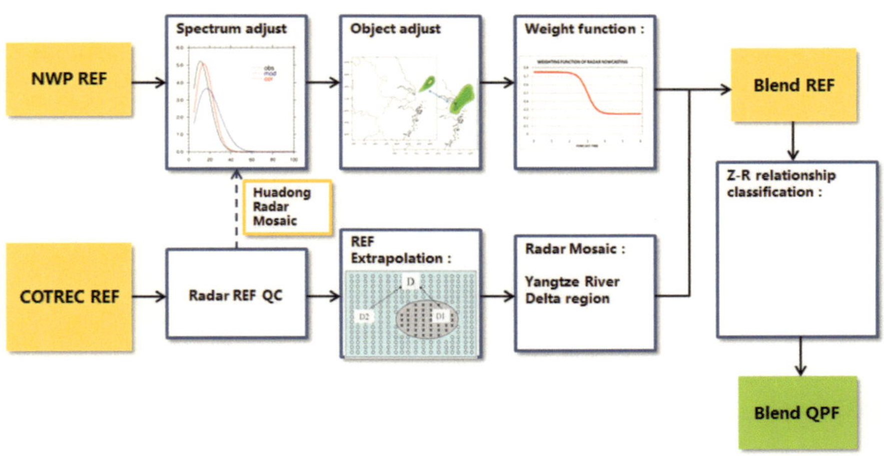

Figure 1. Flowchart of the blending forecast.

is obtained by applying non-convergence algorithm to the TREC field. Finally, by applying the COTREC wind field to the observation radar reflectivity of t2, the extrapolation at t2 + Δt is completed (**Figure 2**).

Traditional COTREC extrapolation is effective in about 2 h for a single radar observation. In order to extend the extrapolation time to more than 2 h, on the one hand, it is necessary to enlarge the radar data and extend the original single radar observation to the radar network over Yangtze River Delta region surrounding Shanghai. Thanks to the East China regional data sharing line established during the 2010 Shanghai World Expo, observations of 11 Doppler radars around Shanghai are available for the needs of longer time quality control and extrapolation (**Table 1**). Furthermore, as theoretically the vector field calculated by the TREC method is only effective for the current time, TREC method tends to lose validity due to the evolution of atmosphere, so when forecast time comes to over 2 h, the average wind field of numerical forecast is gradually taking part to replace the TREC wind field so as to indicate the evolution of the wind field.

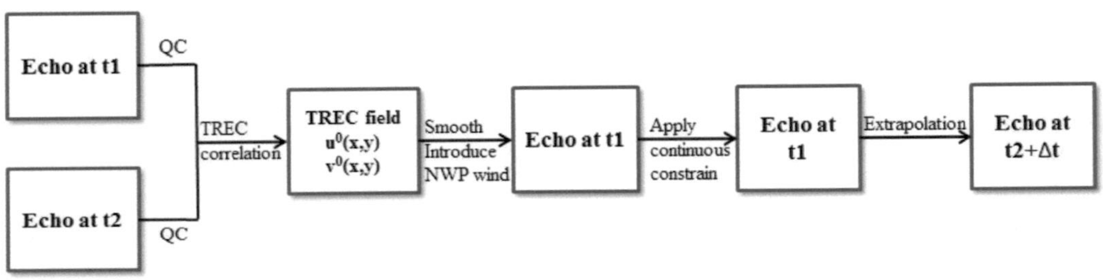

Figure 2. Flowchart of the COTREC extrapolation.

Name	ID	Type	Longitude	Latitude	Antenna altitude (m)
Qingpu	Z9002	SA	120°57′32″	31°4′30″	42.1
Shanghai	Z9210	88D	121°53′05″	31°00′05″	44.2
Nanjing	Z9250	SA	118°41′52″	32°11′27″	138.8
Nantong	Z9513	SA	120°58′33″	32°04′33″	29
Yancheng	Z9515	SA	120°12′07″	33°25′55″	28.3
Hangzhou	Z9571	SA	120°20′15″	30°16′25″	96
Jinhua	Z9579	SB	119°38′39″	29°12′55″	1191.5
Zhoushan	Z9580	SB	122°06′32″	30°04′08″	438.3
Hefei	Z9551	SA	117°15′28″	31°52′1″	165.5
Huangshan	Z9559	SA	118°09′	30°08′	1847
Nanchang	Z9791	SA	115°53′56″	28°35′27″	94.9

Table 1. Radar sites around Shanghai used for extrapolation.

2.2. NWP configuration for blending

For the NWP system, how to configure the numerical model system to make it meet the needs of short-term precipitation forecast as well as the afterward reflectivity blending is a prior task throughout the system configuration. Some of the key issues here include (1) the need to run a regional NWP model with high resolution to resolve weather processes involving small to mesoscale convections, (2) configuration of parametrization schemes in convection scale to keep strength and development of simulated echo system in basically the same range of the radar observation to be assimilated, (3) reduction of spin-up time of NWP to about 1 h to meet the needs for rapid updated short-term prediction, and (4), finally, as traditional short-term precipitation forecast relies heavily on radar observations, it is a common thought that better radar reflectivity assimilation and prediction are the keys to a better NWP for the blending. In the following part of this section, we focus on the NWP system configuration and radar data assimilation to describe their influence on the prediction of radar reflectivity.

2.2.1. The regional NWP system

Research and practice have shown that model spin-up can be effectively reduced by using Rapid Update Cycle [15–18]. By doing RUC is basically to update numerical forecast of the previous time by continuously adding the latest observation data assimilation, thus to reduce spin-up time, and improve the short-term forecast. In this chapter, the Weather Research and Forecasting Model (WRF-ARW, V3.9) and Community Gridpoint Statistical Interpolation (GSI, V3.6) system were chosen as the numerical model and data assimilation system. In order to integrate with the radar extrapolation, the model configuration was focused on the forecast of reflectivity. Current 12 h forecast configuration includes a model grid of 700*700*51, horizontal resolution of 3 km, Thompson's microphysics parameterization, Noah land surface model, RRTM longwave radiation scheme, and Dudhia shortwave radiation scheme. Referring to the short-term system Rap [19] of the National Oceanic and Atmospheric Administration (NOAA), the overall operation scheme was configured to a partially RUC CYCLE run accompanied with a 6 h catch-up CYCLE over the East China (**Figure 3**). GFS forecast from the National Centers for Environmental Prediction (NCEP) was used to provide initial and boundary conditions twice a day at 02/14 BT for the catch-up cycle. The ECFLOW software package developed by the European Centre for Medium-Range Weather Forecasts (ECMWF) is used for task scheduling and monitoring (https://software.ecmwf.int/wiki/display/ECFLOW/). By using such a cycle-updated configuration with conventional data assimilation, spin-up time was essentially eliminated even in the absence of radar data; well-structured precipitation was usually able to be constructed within 1 h.

Cumulus and boundary layer parameterization are two other important schemes for model precipitation. Cumulus parameterization directly influences the thermal environment and precipitation forecast of model [20, 21]. While being the place where most heat and water vapor concentrate in the atmosphere, eddies in the boundary layer were the main trigger mechanism for precipitation, especially in the daytime [22, 23]. The chosen of different boundary layer parameterization schemes would have important implications for forecasting precipitation [24–27]. In the current system, the UW boundary layer scheme [28] and the GF cumulus

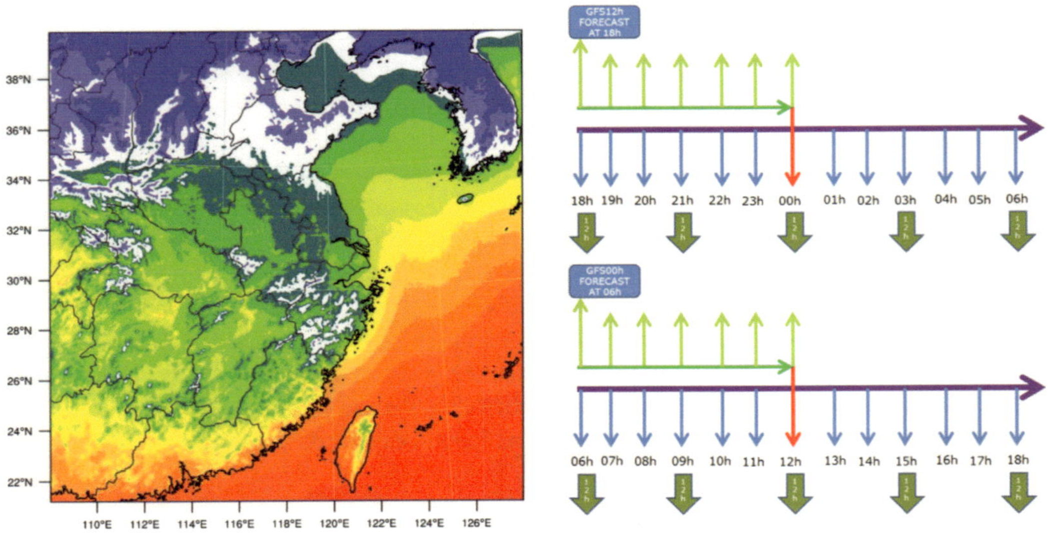

Figure 3. Model domain and configuration of the rapid update cycle (blue arrows) and the catch-up cycle (green arrows).

parameterization scheme [29] are chosen to be the boundary layer and cumulus parameterization scheme. Cumulus convection and boundary layer parameterization schemes can help the establishment of rain bands within the NWP model, while for our experiment, the selection of parametrization schemes should also be considered along with the adaptability of radar assimilation and extrapolation. With the application of radar data and cycled start-up, precipitation should already form in model start for most cases. The maintenance and development of the initial rain system are more important here. Research has found that the UW scheme showed a stronger mixing of the boundary layer turbulence in the daytime [30] and thus has a good inhibition on the afternoon false alarm of local convection, which is usually the case for regional NWPs. On the cumulus parameterization, the GF scheme was chosen for two reasons: one is that it can be applied to high-resolution numerical models [31] and the other is for the optional shallow convection scheme with GF scheme, which can also help to suppress false convection by distinguishing local convection type.

2.2.2. Radar data assimilation for reflectivity

Currently, weather radar is the only way to a continuous, high-resolution, three-dimensional direct observation of convective scale precipitation. Conventional Doppler weather radar observes radial wind and reflectivity, if one wants to combine them into the initialization of NWP; the key point here is to make radar observation compatible with the thermal dynamic fields of model atmosphere, so that they are neither smoothed away as noises nor to induce false development to the model. By introducing radar radial wind and reflectivity data, Xiao et al. [32] used the MM5 3DVAR system to improve initial wind field and moisture distribution of model, thus improving short-term precipitation simulation. Xue et al. [33] assimilated radar reflectivity data within a time window. By using the ARPS assimilation system, they were able to retrieve the distribution of water vapor and precipitation particles in the area of radar observation and thus to improve the initial condition of model and afterward tornado simulation. Some studies found that [34–36] the assimilation of radial wind data was generally less important to precipitation than reflectivity data.

The assimilation of radar reflectivity at present was usually done through "diabetic initialization," which is to convert the observation of radar reflectivity to the model variable such as cloud and precipitation particles. At the same time, in-cloud atmosphere profile was adjusted according to latent heat release and water vapor saturation ratio [37]. Researches have shown that these kinds of assimilation can effectively reduce NWP spin-up time and improve forecast from the model start to a few hours later. While in the case of radial wind, conventional variation method (such as 3D-/4D-VAR) was used to do the assimilation. It should be pointed out that for the assimilation of single radar radial wind, the improvement to the NWP's three-dimensional wind field was often limited, and it was usually done along with the assimilation of reflectivity to get a better improvement.

In recent years, China made a great effort in the construction of various meteorological observation networks. With the construction of the China's next-generation weather radar (CINRAD) network, even trivial precipitation is now hard to miss from observation. Especially to the east part of China, where radar observations were full covered, the integrated radar reflectivity observation can basically be viewed as the real-time precipitation distribution. After data quality control and interpolation, the complex cloud analysis scheme in GSI system was carried out to assimilate the reflectivity data. Model profile and particle mass variables were thus moderated to contain the observed precipitation system.

As shown in **Figure 4**, the reflectivity of 51 Doppler radars in the middle and east part of China was used. The new-generation weather radar 3D mosaic system [38, 39] of the Chinese Academy of Meteorological Sciences (CAMS) was used to integrate the raw data into a three-dimensional mosaic reflectivity. The system was used to apply simple quality control to single radar observation and convert it to three-dimensional Cartesian coordinates. Afterward, data from each radar in the region was integrated together to the three-dimensional reflectivity mosaic. As GSI needed mosaic data of 31 levels which was a bit different from CinradMosaic, another vertical interpolation to the mosaic data was applied. Finally, the mosaic data was converted into Bufr formats onto the horizontal grid of NWP using the GSI radar data interface, so to be used in the cloud analysis [40, 41].

Due to the fine coverage of the radar network observation, the distribution of precipitation can be basically indicated at model initial time through assimilation. In order to perform a better blending with the radar extrapolation afterward, the NWP system also needs to retain as much information as possible from the initial radar assimilation. If the difference between observation and prediction was so large that one can hardly relate one to the other, trying to blend them would certainly be difficult. Therefore, before the assimilation of reflectivity, all the initial precipitation particles from the original background were removed, so that the simulated reflectivity of model after assimilation was consistent with the observation (**Figure 4**). Although

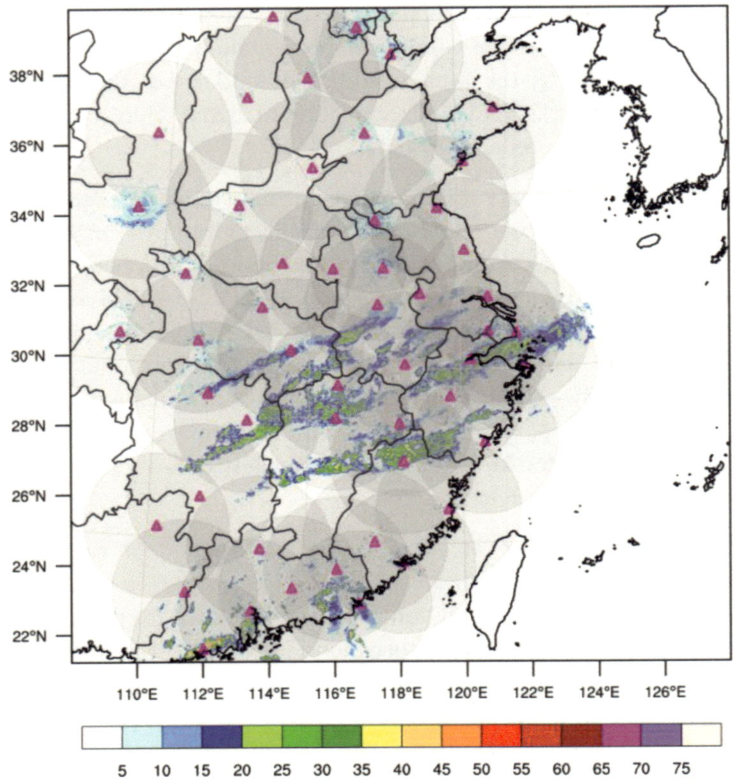

Figure 4. Distribution of radar stations (purple triangles) used in assimilation and retrieved initial composite reflectivity (shaded) from assimilation at 11:00 on June 5, 2018.

eliminating previous particles could reduce false precipitation to a certain extent, due to the application of hourly update cycle, radar observations were continuously added into the NWP. Adjustments done by cloud analysis could still easily lead to overloading of the NWP's hydrometer fields, which further caused the overdevelopment of precipitation. One way to prevent this problem in RAP was to reduce the overstatement of precipitation by setting up a threshold to limit the hydrometer change in every cycle. Similarly in this experiment, assimilation of reflectivity was restrained in two aspects. Firstly, the assimilation reflectivity was performed only in the region where reflectivity was above 15dBZ. Secondly, the retrieved hydrometer mass field was restricted to the limit of 3 g/kg in high-reflectivity area and 1 g/kg in the low-reflectivity area. **Figure 5** provides a comparison of the 3–9 h radar composite reflectivity forecast and observation for a rainfall event in Central China on April 23, 2018. As shown in the figure, the system could maintain radar reflectivity observation from assimilation and continue to forecast its development up to about 9 h. Meanwhile, the overall intensity and position of the simulated reflectivity are kept consistent with the observation. In this way, both the prediction and observation had a good similarity in intensity and shape; then, it is possible to calibrate the forecast echoes based on the comparison of the two.

http://dx.doi.org/10.5772/intechopen.81632

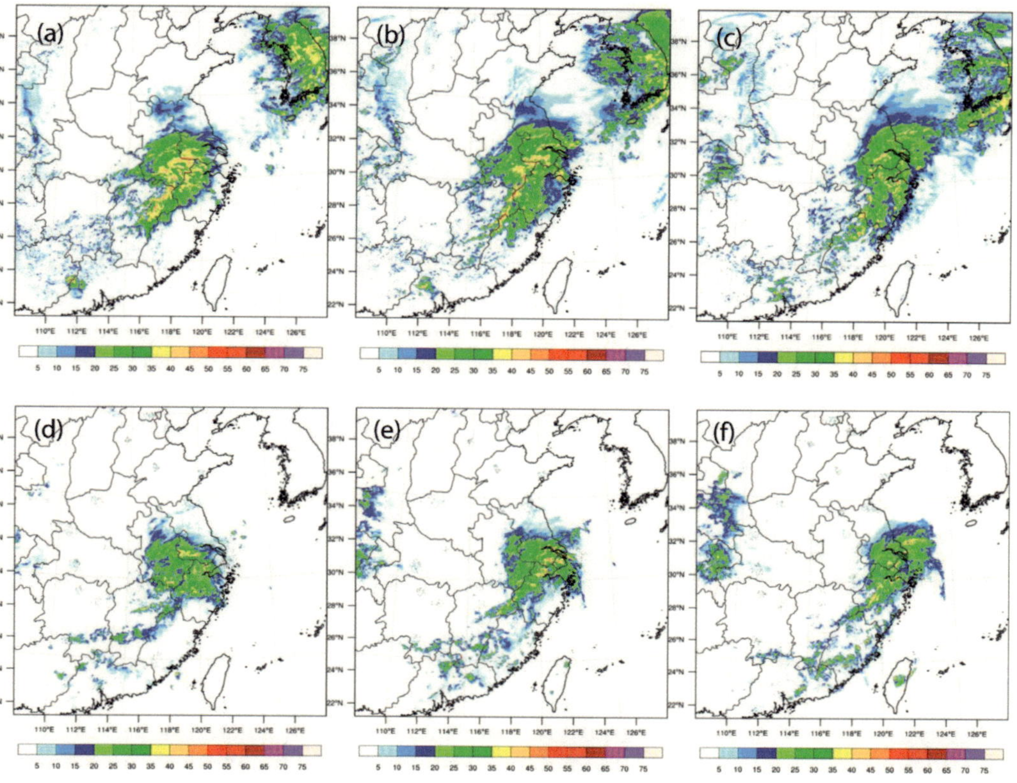

Figure 5. Composite reflectivity of forecast (a/b/c) and observation (d/e/f). Forecast started on 08:00 23rd April, 2018. (a)(d) 11:00 BT, (b)(e) 15:00 BT, and (c)(f) 17:00 BT.

2.3. Calibration to the NWP forecasted reflectivity

While it often took the short-term NWP forecast a delay of about 2 h to do data collection and time integral, so even by the time forecasters got the NWP forecast, there would already be errors on intensity and position. In order to integrate the NWP forecast with radar extrapolation, it is necessary to calibrate the NWP prediction according to the current observation before it can be integrated with the extrapolation. The main calibrations here include:

1. Calibration of the intensity. A Weibull function was used here to fit the probability distribution of forecast and observation reflectivity separately. The probability density function is:

$$f(x;\lambda,k)=\begin{cases}\frac{k}{\lambda}\left(\frac{x}{\lambda}\right)^{k-1}e^{-(x/\lambda)^{k}}, & x\ge 0\\ 0, & x<0\end{cases}$$

where k is the shape parameter and λ is the scale parameter; they were used to determine the resemblance of forecast and observation. If the two sets of parameters deviated from above a certain threshold, calibration was carried out by doing a proportional scaling to the forecast reflectivity, so that the distribution of k and λ was closer to the observation ones. Recursive scaling was carried out until the final parameters meet the threshold (**Figure 6**). The same scaling value was then preserved and applied to the forecasts afterward but should be slightly weakened as forecast hour increased.

2. Calibration of the position. Position error of NWP was evaluated based on a target recognition algorithm of SCMO. Firstly, a low-pass filtering was carried out to remove noise and singularity for both forecast and observation. Target recognition was then carried out by identifying different blocks of distinct echoes. After target recognition, the predicted echoes were associated with the observation target by target by comparing the deviation characteristics such as center of gravity, scale, shape, period, intensity, and so on. Finally, when two targets from forecast and observation were identified as the same echo, the displacements of their centers of gravity were used to adjust the position of the original forecast reflectivity (**Figure 7**).

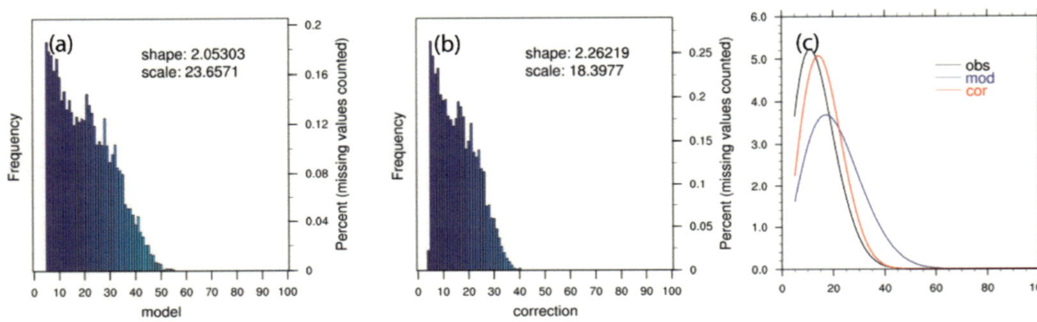

Figure 6. Intensity adjustment for a 3 hour forecast started at 11:00 BT 20th April, 2016. (a) NWP prediction (original), (b) NWP forecast (after calibration), and (c) Weibull fit for the forecast before/after calibration and the observation.

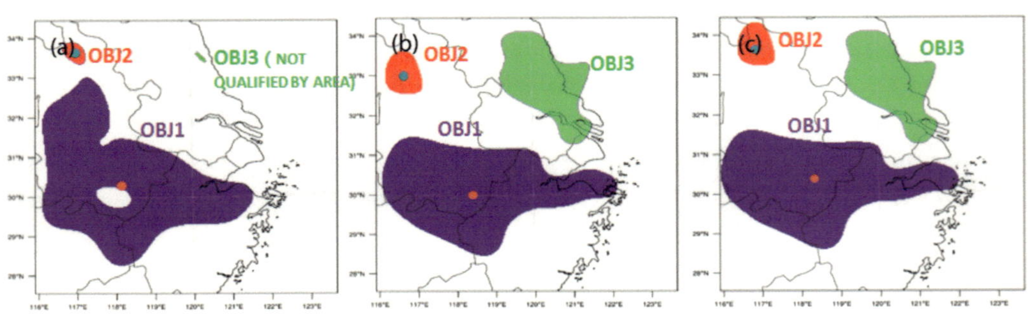

Figure 7. Position adjustment for a 3 hour forecast started at 11:00 BT 20th April, 2016. (a) target identification for observation (after low pass filtering), (b) target identification for NWP prediction (after low-pass filtering), and (c) target identification for NWP prediction (after target adjustment).

2.4. Weighted blending and Z-R relation determination

Different weight coefficients were applied to the extrapolation and calibrated NWP prediction in the stage of reflectivity blending. Coefficients were designed basically to make the extrapolation dominant in the early time (0–2 h) and make the NWP prediction smoothly take control later on (4–6 h). The weight function here used was based on the same equation used by Chen et al. [13]:

$$W_{NWP} = 0.5 + 0.5*(1 + \tanh(t - t0))$$

$$W_{EX} = 1 - W_{NWP}$$

Precipitation type	Z-R coefficients	
	A	B
Convective precipitation	200	1.46
Mixed precipitation	254	1.33
Stable precipitation	263	1.26
Tropical precipitation	200	1.20
Default	300	1.4

Table 2. Coefficients of the five Z-R relation functions in operation at SCMO.

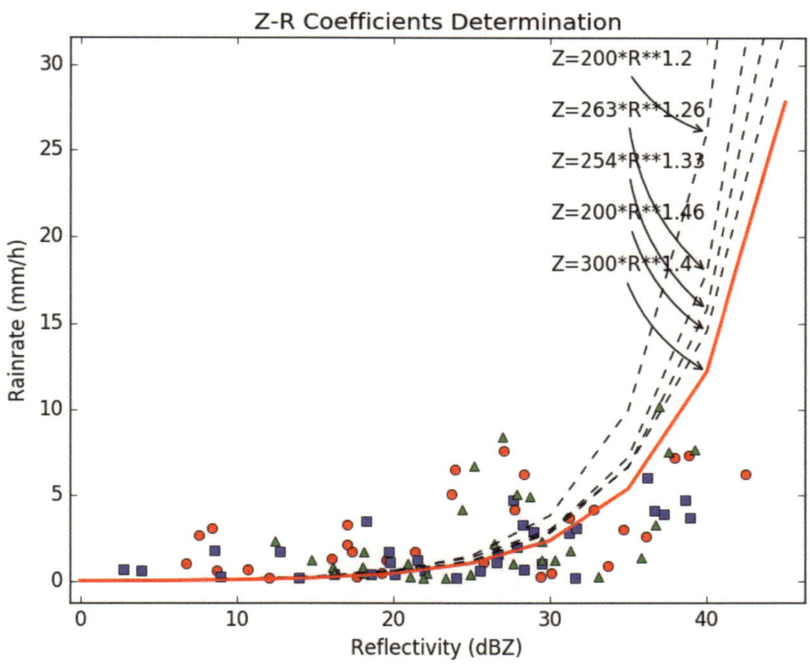

Figure 8. Real-time Z-R relation function determination.

where t0 is the time when the extrapolation weight took the value of 0.5. At present the value of t0 is adjusted within 3–4 according to the scale of the system empirically. The larger the extrapolation echo scale was, the greater the t0 value would be.

Finally, the blended radar reflectivity is converted to precipitation rate through a Z-R relation function:

$$Z = AR^{B}$$

There were five Z-R functions right now in operation at SCMO (**Table 2**). In order to decide which one to use in the final conversion, a real-time evaluation for each function was applied. Precipitations from the ground AWS observation along with their observed reflectivity in the latest 3 h were used as statistical samples, mean absolute errors of these samples for each Z-R function was calculated, and the one with the smallest error was selected to use (**Figure 8**). The final operation run was a 6-h blending QPF forecast updated every 30 min with 3 km/10 min space-time resolution.

3. Forecast case and evaluation

The forecast to a rainfall event occurred in the late evening of July 5, 2018, is exhibited below for demonstration. This event happened during the Meiyu season in the southern part of

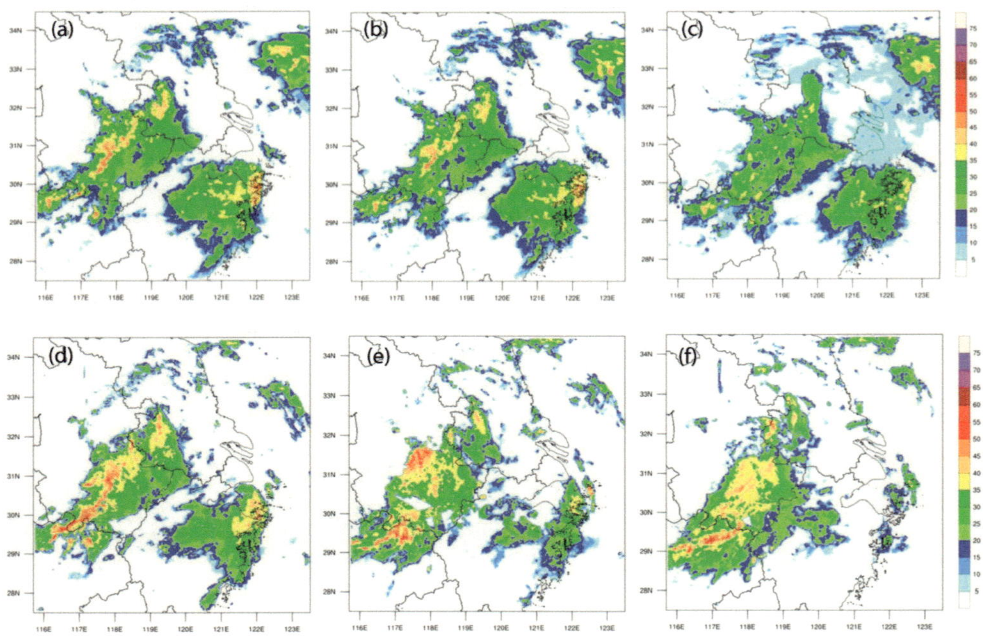

Figure 9. Composite reflectivity forecast and observation at 22:00/23:00/24:00 BT 5th July, 2018. (a)(b)(c) Blending forecast and (d)(e)(f) Mosaic observation.

Jiangsu and Anhui provinces. The rainfall was produced by a linear mesoscale convective system (MCS) which occurred on the low-level vortex of the Meiyu front. Early convections formed in the afternoon of July 5 in the middle of Anhui province and then moved southeast. The precipitation lasted to until 18:00 BT in the evening, and new convections started to form to the southwest side of the former convection in the southwest of Anhui. The convections then organized into a quasi-linear convection zone, which lead to the heavy rainfall in the wide of south Anhui. **Figures 9, 10** provide composite reflectivity evolution to the process of both the blending forecast and radar mosaic observation. Forecast started at 21:00 BT July 5 when the linear structure of the convection system was basically formed and about to enhance. **Figure 9** shows that the first 3 h forecasts of the blending are mainly the results of extrapolation. The echo in south Anhui moves southeast while sustaining the initial intensity. Compared with the observation, the shape of the forecast echo was still satisfying, but the intensity basically keeps the same as initial. The influence of NWP is gradually strengthened after 3 h (**Figure 10**). New convection develops to the bordering area of Zhejiang, Jiangxi, and Anhui while echoes to the bordering area of Jiangsu and Zhejiang. While the moving direction of original extrapolation echo is mainly southeast, the 6 h blending forecast overall shows a rainfall system moving south and developing, which is consistent with the observation. Therefore, the blending forecast shows a better effect compared to that of extrapolation and NWP prediction at 0–6 h. There is also precipitation in the east of Zhejiang at the beginning of the forecast, which was mainly generated by convections in the mountainous area of Zhejiang merging together.

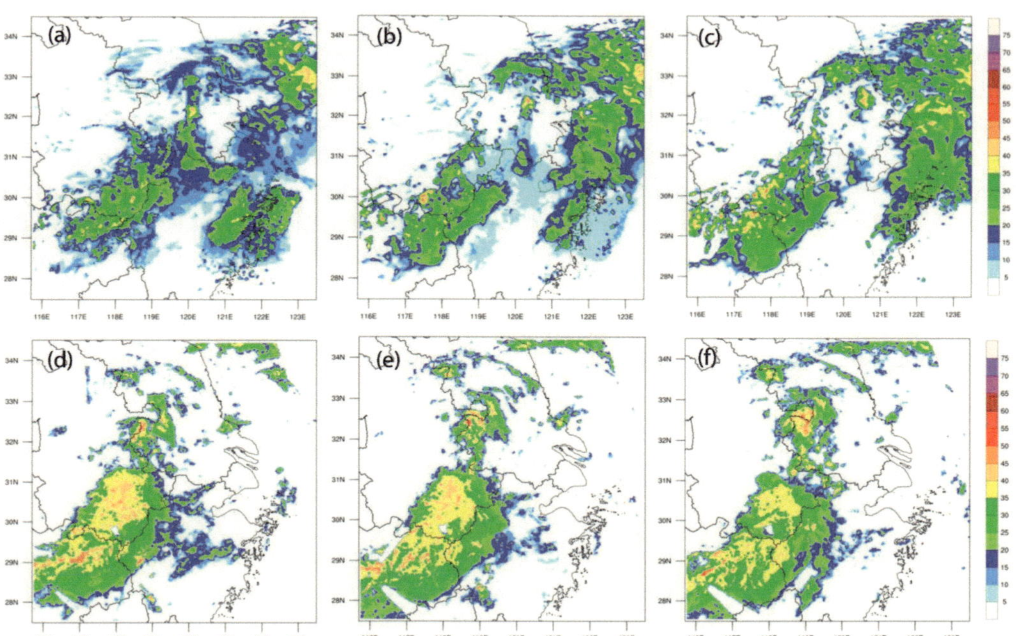

Figure 10. Composite reflectivity forecast and observation at 01:00/02:00/03:00 BT 6th July, 2018. (a)(b)(c) Blending forecast and (d)(e)(f) Mosaic observation.

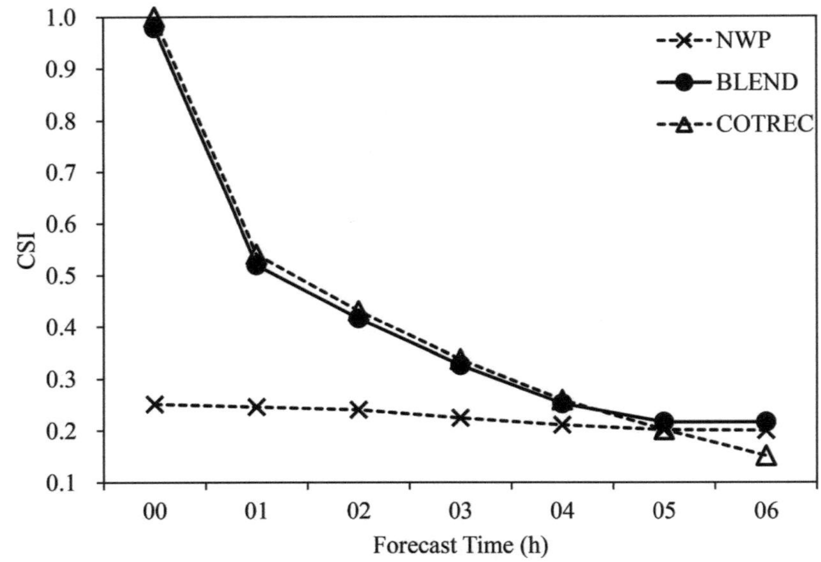

Figure 11. Critical success index for 94 forecast experiments of four rainfall events in July 2018 (threshold: 20dBZ).

It gradually disperses while moving east after late afternoon, which is also the case in the blending forecast.

To objectively illustrate the effects of blending forecast, experiments were carried out for four rainfall events in the later part of Meiyu period from July 1–9, 2018, for forecast examination. With a total of 94 forecasts, the critical success index (CSI) for the reflectivity threshold of 20 dBZ is shown in **Figure 11**. The forecast score of extrapolation overrides that of the NWP in a longer time than common sense. The NWP only takes over at forecast time after 4 h. Although it may be case dependent, the extrapolation of the mosaic reflectivity does have its validity in longer forecast times in some occasion. And due to the differential weight coefficient adjustment, the blending forecast is able to keep resembling to that of extrapolation within 4 h overmatching the NWP. When the extrapolation score falls below the NWP after 4 h, weight of the NWP forecast begins to dominate. The fact that blending forecast scores override the NWP scores in all times shows the positive effect of NWP calibration to the forecasted reflectivity.

4. Conclusions and discussions

This chapter introduced the 0–6 h blending QPF forecast in SCMO, especially the configuration and calibration of NWP in it. The forecast is based on an extended extrapolation of 11 Doppler radars over the Yangtze River Delta and reflectivity forecast of a high-resolution rapid updated NWP system. Intensity correction based on Weibull transform, position correction based on target recognition and matching, weight coefficient adjustment based on precipitation scale, and Z-R relationship determination based on real-time AWS observation are used to calibrate and blend the NWP forecast.

1. Extrapolation time can be extended when combining observations of multiple radars in the region and using NWP wind forecast to do the extrapolation when forecast time increases.

2. As NWP is still not perfect for all the forecast elements, configuration of the NWP system is better done according to one's special purpose of use. For short-term efficiency of the reflectivity forecast, attention needs to be paid to decrease model spin-up, improving radar data assimilation and various parameterization schemes. Based on the choice of the UW boundary layer scheme, the GF cumulus scheme with shallow convection, complex cloud analysis of three-dimensional reflectivity, and rapid update cycle along with catch-up configuration, the NWP system is able to sustain initial observation echo and forecast its development in the following hours and, at the same time, keep a good reduction to false alarms of model convections.

3. Calibrations to the NWP forecasts are carried out based on real-time error analysis before they can be integrated with the extrapolation. Spectrum distribution of forecast reflectivity is adjusted by the standards made by a Weibull distribution fitting. Echo positions are adjusted by target identification and matching according to parameters such as weight, scale, and strength. The final QPF is achieved by real-time Z-R relation determination. Forecast experiments were carried out, and results show that blending extrapolation and the NWP prediction can get a better short-term prediction of 0–6 h with the advantages of both.

4. Some of the technical scheme used in this experiment is still immature. For example, for the radar radial wind assimilation, although in some cases it might have got some improvement to the movement and intensity of precipitation, but due to the data coverless and lack of quality control, it was common that there were severe boundary effects which contaminated forecast later that we had to exclude it in the system. For the extrapolation, some state-of-the-art machine learning algorithms can already do forecasts with echo intensity development. And for the areas without radar coverage, how to use satellite data to get a 3D precipitation observation as a substitution is sure to be key to get a better short-term forecast there.

Author details

Hai Chu*, Mengjuan Liu, Min Sun and Lei Chen

*Address all correspondence to: chhaui@163.com

Shanghai Central Meteorological Observatory, Shanghai, China

References

[1] World Meteorological Organization World Weather Research Programme. Strategic Plan for the Implementation of WMO's World Weather Research Programme (WWRP): 2009–2017; 2009. 37pp. Available from: http://www.wmo.int/pages/prog/arep/wwrp/new/documents/final_WWRP_SP_6_Oct.pdf

[2] Sun J, Xue M, Wilson JW, et al. Use of NWP for nowcasting convective precipitation: Recent progress and challenges. Bulletin of the American Meteorological Society. 2014;**95**: 409-426

[3] Developmental Testbed Center. Gridpoint Statistical Interpolation Advanced User's Guide Version 3.5.0.0; 2016. Available from: http://www.dtcenter.org/com-GSI/users/docs/index.php. 143 pp

[4] Dixon M, Wiener G. TITAN: Thunderstorm identification, tracking, analysis, and nowcasting—A radar-based methodology. Journal of Atmospheric and Oceanic Technology. 1993;**10**(10):785

[5] Rinehart RE, Garvey ET. Three-dimensional storm motion detection by conventional weather radar. Nature. 1978;**273**(5660):287-289

[6] Wilson JW, Crook NA, Mueller CK, et al. Nowcasting thunderstorms: A status report. Bulletin of the American Meteorological Society. 1998;**79**(10):2079-2100

[7] Gibson JJ. The Ecological Approach to Visual Perception. Houghton Mifflin; 1979

[8] Yilmaz A. Object tracking: A survey. ACM Computing Surveys. 2006;**38**(4):13

[9] Lei C, Jianhua D, Lan T. Application of an improved TREC algorithm (COTREC) for precipitation nowcast. Journal of Tropical Meteorology (in Chinese). 2009;**25**(1):117-122

[10] Nimrod GBW. A system for generating automated very short range forecasts. Meteorological Applications. 1998;**5**(1):1-16

[11] Wilson J, Feng Y, Chen M, et al. Nowcasting challenges during the Beijing Olympics: Successes, failures, and implications for future nowcasting systems. Weather and Forecasting. 2010;**25**(6):1691-1714

[12] Li PW, Wong WK, Chan KY, et al. SWIRLS–An evolving nowcasting system. Hong Kong Observatory Technical Note No. 100; 2000

[13] Conglan C, Mingxuan C, Jianjie W, Feng G, Yeung Linus HY. Short-term quantitative precipitation forecast experiments based on blending of nowcasting with weather prediction. Acta Meteorologica Sinica (in Chinese). 2013;**71**(3):397-415

[14] Feng X, Xing W, Shuang W, Chunsheng M, Linhan Z. Dynamic blending probabilistic precipitation forecast method based on radar extrapolation and numerical weather prediction. Meteorological Science and Technology (in Chinese). 2017;**45**(6):1036-1042

[15] Benjamin SG, Dévényi D, Weygandt SS, et al. An hourly assimilation forecast cycle: The RUC. Monthly Weather Review. 2004;**132**(2):495

[16] Alexander C, Weygandt S, Benjamin SG, Smirnova TG, Brown JM, Hofmann P, et al. The high resolution rapid refresh (HRRR): Recent and future enhancements, time-lagged ensembling, and 2010 forecast evaluation activities. In: Proceedings of the 24th Conference on Weather and Forecasting/20th Conference on Numerical Weather Prediction. Seattle, WA: The American Meteor Society; 2011. 12B.2. Available from: https://ams.confex.com/ams/91Annual/webprogram/Paper183065.html

[17] Min C, Shuiyong F, Jiqin Z, Yanli C, Zheng Z, Chaolin Z, et al. An experimental study of assimilating the global position system-precipitable water vapor observations into the rapid updated cycle system for the Beijing area. Acta Meteorologica Sinica (in Chinese). 2010;**68**(4):450-463

[18] Baode C, Wang X, Hong L, Lei Z. An overview of the key techniques in rapid refresh assimilation and forecast. Advances in Meteorological Science and Technology (in Chinese). 2013;**3**(02):29-35

[19] Benjamin SG, Coauthors. A north American hourly assimilation and model forecast cycle: The rapid refresh. Monthly Weather Review. 2016;**144**:1669-1694

[20] Emori S, Nozawa T, Numaguti A, et al. Importance of cumulus parameterization for precipitation simulation over East Asia in June. Journal of the Meteorological Society of Japan. 2001;**79**(4):939-947

[21] Anning H, Yaocun Z, Jian Z. Sensitivity of simulation of different intensity of summer precipitation over China to different cumulus convection parameterization schemes. Chinese Journal of Atmospheric Sciences (in Chinese). 2009;**33**(6):1212-1224

[22] Wilson JW, Schreiber WE. Initiation of convective storms at radar-observed boundary-layer convergence lines. Monthly Weather Review. 1986;**114**:2516-2536

[23] Wilson JW, Roberts RD. Summary of convective storm initiation and evolution during IHOP: Observational and modeling perspective. Monthly Weather Review. 2006;**134**:23-47

[24] Cohen AE, Cavallo SM, Coniglio MC, et al. A review of planetary boundary layer parameterization schemes and their sensitivity in simulating southeastern U.S. cold season severe weather environments. Weather and Forecasting. 2015;**30**(3):150224120634008

[25] Hong SY, Pan HL. Nonlocal boundary layer vertical diffusion in a medium-range forecast model. Monthly Weather Review. 1996;**124**(124):2322

[26] Li X, Pu Z. Sensitivity of numerical simulation of early rapid intensification of hurricane Emily (2005) to cloud microphysical and planetary boundary layer parameterizations. Monthly Weather Review. 2015;**136**(136):4819

[27] Ma L, Bao X. Research progress on physical parameterization schemes in numerical weather prediction models. Advances in Earth Science. 2017;**32**(7):679-687

[28] Bretherton CS, Park SS. A new moist turbulence parameterization in the community atmosphere model. Journal of Climate. 2009;**22**(12):3422-3448

[29] Grell GA, Freitas SR. A scale and aerosol aware stochastic convective parameterization for weather and air quality modeling. Atmospheric Chemistry and Physics Discussions. 2013; **13**(9):23845-23893

[30] Wenyan H, Xinyong S, Weiguo W, Wei H. Comparison of the thermal and dynamic structural characteristics in boundary layer with different boundary layer parameterizations. Chinese Journal of Geophysics (in Chinese). 2014;**57**(5):1399-1414

[31] Zhe X. Impact of different convective parameterization on simulation of precipitation for the Heihe River basin. Advances in Earth Science (in Chinese). 2014;**29**(5):590-597

[32] Xiao Q, Kuo YH, Sun J, et al. Assimilation of doppler radar observations with a regional 3DVAR system: Impact of Doppler velocities on forecasts of a heavy rainfall case. Journal of Applied Meteorology. 2005;**44**(6):768-788

[33] Xue M, Wang D, Gao J, Brewster K, et al. The advanced regional prediction system (ARPS), storm-scale numerical weather prediction and data assimilation. Meteorology and Atmospheric Physics. 2003;**82**(1):139-170

[34] Fiedler BH. A wind transform for acoustic adjustment in compressible models. Monthly Weather Review. 2010;**130**(3):741

[35] Hu M, Xue M, Brewster K. 3DVAR and cloud analysis with WSR-88D level-II data for the prediction of the Fort Worth, Texas, tornadic thunderstorms. Part I: Cloud analysis and its impact. Monthly Weather Review. 2006;**134**(2):699-721

[36] Hu M, Xue M, Gao J, Brewster K. 3DVAR and cloud analysis with WSR-88D level-II data for the prediction of the Fort Worth, Texas, Tornadic Thunderstorms. Part II: Impact of radial velocity analysis via 3DVAR. Monthly Weather Review. 2006;**134**:699-721

[37] Zhang J. Moisture and diabatic initialization based on radar and satellite observations [Ph. D. dissertation]. University of Oklahoma; 1999

[38] Yanjiao X, Liping L. Study of methods for interpolating data from weather radar network to 3-D grid and mosaics. Acta Meteorologica Sinica. 2006;**64**(5):647-656

[39] Zhiqiang Z, Liping L, Xie M, Wang H, Yanjiao X. A display system of CINRAD 3D mosaic products. Meteorological Monthly (in Chinese). 2007;**33**(9):19-24

[40] Shao H, Derber J, Huang XY, Hu M, Newman K, Stark D, et al. Bridging research to operations transitions: Status and plans of community GSI. Bulletin of the American Meteorological Society. 2016;**97**:1427-1440

[41] Hu M, Shao H, Stark D, Newman K, Zhou C, Ge G, et al. Grid-Point Statistical Interpolation (GSI) User's Guide Version 3.6; 2017. Developmental Testbed Center. Available from: http://www.dtcenter.org/com-GSI/users/docs/index.php. 149 pp

Chapter 5

Spectral Representation of Time and Physical Parameters in Numerical Weather Prediction

Kristoffer Lindvall and Jan Scheffel

Additional information is available at the end of the chapter

http://dx.doi.org/10.5772/intechopen.80351

Abstract

Numerical weather prediction (NWP) is a difficult task in chaotic dynamical regimes because of the strong sensitivity to initial conditions and physical parameters. As a result, high numerical accuracy is usually necessary. In this chapter, an accurate and efficient alternative to the traditional time stepping solution methods is presented; the time-spectral method. The generalized weighted residual method (GWRM) solves systems of nonlinear ODEs and PDEs using a spectral representation of time. Not being subject to CFL-like criteria, the GWRM typically employs time intervals two orders of magnitude larger than those of time-stepping methods. As an example, efficient solution of the chaotic Lorenz 1984 equations is demonstrated. The results indicate that the method has strong potential for NWP. Furthermore, employing spectral representations of physical parameters and initial values, families of solutions are obtained in a single computation. Thus, the GWRM is conveniently used for studies of system parameter dependency and initial condition error growth in NWP.

Keywords: NWP, time-spectral, chaotic, error analysis, initial condition

1. Introduction

Numerical methods are routinely used when modeling complex systems such as meteorological and atmospheric systems. These systems can be described by nonlinear ordinary and partial differential equations. Nonlinear systems in particular, as opposed to linear systems, constitute specific numerical challenges in terms of processing and memory requirements.

In order to efficiently use the computer's resources, it is important to choose a suitable numerical method [1].

Typically, when solving initial value ODEs, it is common practice to use finite difference methods (FDM) that discretize the temporal domain into finite steps. Information from the initial state and previously computed steps are used to estimate the solution at a "future" discrete time step. This is an intuitive and powerful method because we can easily visualize the cause and effect of each step. However, the problem with these techniques when solving complex systems is that small temporal steps are required in order for a solution to be found. In order to acquire a stable solution, the Courant-Friedrichs-Lewy (CFL) condition needs to be satisfied when employing more accurate, explicit FDM, thus limiting time step length.

Another approach is to use a time-spectral method, which takes a bird's eye view of the problem [2–7]. Instead of focusing on small local steps in time, an approximate solution that could fit the entire temporal domain is postulated. This approximate solution is set in the form of a finite series of basis functions. Thus, the unknowns of the equations are thereby changed from the physical variables to the coefficients of the solution ansatz. The procedure allows for highly accurate and global temporal solutions.

In the following, we will present application of the GWRM [7] to the Lorenz 1984 [8] differential equations. This is a set of three coupled, nonlinear, and chaotic ODEs, providing a basic model of the important process of Hadley tropical atmospheric circulation. In Section 2, Method, we present a short derivation of the GWRM. Section 3 will introduce the Lorenz equations and how they can be solved by the time-spectral method. In this section, we also show how the parameter dependence of the solution can be found from a *single* GWRM computation. We also show that, similarly, the dependence on initial conditions can be found from one computation only. A discussion is given in Section 4, followed by conclusion in Section 5.

2. Method

The time-spectral method employed here is the generalized weighted residual method (GWRM). The physical equations of this Galerkin method are projected onto a space of weighted orthogonal basis functions resulting in a set of algebraic equations. The residual of the differential equation is necessarily zero for the exact solution, thus the numerical method seeks to solve for the coefficients that minimize the residual in the interval where orthogonality holds.

In this procedure, all dimensions can be included, thus the term "generalized." This means that the dependence on all *temporal*, *spatial*, and *parameter* domains can be included in the algebraic equations.

A key characteristic of Chebyshev polynomials, and a main reason for choosing them as basis functions, is their *minimax property* [9]. This property states that a Chebyshev series of order n is the best approximation of the same Chebyshev series of order $n + 1$. The implication is that the highest mode can be neglected without deteriorating the approximate solution. This enables computation of nonlinear products entirely within spectral space. Actually, all GWRM numerical computations are carried out in spectral space. In contrast to, for example, Fourier series, Chebyshev series are also conveniently *real*. Any type of boundary condition can be employed, such as Dirichlet, Neumann, and periodic boundary conditions. Finally, Chebyshev polynomial series are known to converge rapidly when approximating smooth functions.

2.1. Generalized weighted residual method

Consider the following set of differential equations,

$$\frac{\partial \mathbf{u}}{\partial t} - F[\mathbf{u}] = s, \tag{1}$$

where $\mathbf{u}(t, \mathbf{x}; \mathbf{p})$ is the unknown variable with a temporal, spatial, and parameter dependence. F is a linear or nonlinear operator, and s is a source term. The GWRM solution ansatz takes the form, in the case of a singular physical dimension and one parameter,

$$u(t, x; p) \approx \mathrm{U}(\tau, \xi; P) = \sum_{k=0}^{K}{}' \sum_{l=0}^{L}{}' \sum_{m=0}^{M}{}' a_{klm} T_k(\tau) T_l(\xi) T_m(P). \tag{2}$$

Here, $T_n(x) = \cos(n \arccos(x))$ is the Chebyshev polynomial of the first kind, a_{klm} are coefficients, and the prime ($'$) denotes that the first term of each summation should be multiplied by 1/2. Since Chebyshev polynomials are orthogonal in the interval $[-1, 1]$, variable transformations are applied for general intervals,

$$\tau = \frac{t - A_t}{B_t}, \quad \xi = \frac{x - A_x}{B_x}, \quad P = \frac{p - A_p}{B_p}, \tag{3}$$

where $A_z = (z_1 + z_0)/2$ and $B_z = (z_1 - z_0)/2$, z being the variable t, x, and p with their respective upper and lower bounds. The Picard integral is then applied to (1), after which the residual of the differential equation is formed,

$$R = u(t, x; p) - \left[u(t_0, x; p) + \int_{t_0}^{t} (F[u] + s) dt' \right]. \tag{4}$$

The residual is multiplied by weighted Chebyshev polynomials and then integrated over the computational domain, whereafter it is set to zero. The equation takes the form

$$\int_{p_0}^{p_1} \int_{x_0}^{x_1} \int_{t_0}^{t_1} R T_q(\tau) T_r(\xi) T_s(P) \mathrm{w}_t \mathrm{w}_x \mathrm{w}_p dt dx dp = 0, \tag{5}$$

where the Chebyshev weight is $\mathrm{w}_\zeta = \left(1 - \zeta^2\right)^{-1/2}$, ζ being the transform variables τ, ξ, and P. Eq. (5) has the form of an *integral minimum* in calculus of variation, which states that if R is smooth and differentiable, and the basis functions and weights are nonzero, then R will be minimized in the domain [10].

Continuing, we expand Eq. (5),

$$\int_{p_0}^{p_1}\int_{x_0}^{x_1}\int_{t_0}^{t_1}\left\{u(t,x;p)-\left[u(t_0,x;p)+\int_{t_0}^{t}(F[u]+s)dt'\right]\right\}T_q(\tau)T_r(\xi)T_s(P)\mathrm{w}_t\mathrm{w}_x\mathrm{w}_p dtdxdp=0, \quad (6)$$

yielding the following relations for each term of (6):

$$\int_{p_0}^{p_1}\int_{x_0}^{x_1}\int_{t_0}^{t_1}u(t,x;p)T_q(\tau)T_r(\xi)T_s(P)\mathrm{w}_t\mathrm{w}_x\mathrm{w}_p dtdxdp=B_tB_xB_p\left(\frac{\pi}{2}\right)^3 a_{qrs} \quad (7)$$

$$\int_{p_0}^{p_1}\int_{x_0}^{x_1}\int_{t_0}^{t_1}u(t_0,x;p)T_q(\tau)T_r(\xi)T_s(P)\mathrm{w}_t\mathrm{w}_x\mathrm{w}_p dtdxdp=B_tB_xB_p\left(\frac{\pi}{2}\right)^2 \pi\delta_{q0}b_{rs} \quad (8)$$

$$\int_{p_0}^{p_1}\int_{x_0}^{x_1}\int_{t_0}^{t_1}\left(\int_{t_0}^{t}F[u]dt'\right)T_q(\tau)T_r(\xi)T_s(P)\mathrm{w}_t\mathrm{w}_x\mathrm{w}_p dtdxdp=B_tB_xB_p\left(\frac{\pi}{2}\right)^3 A_{qrs} \quad (9)$$

$$\int_{p_0}^{p_1}\int_{x_0}^{x_1}\int_{t_0}^{t_1}\left(\int_{t_0}^{t}sdt'\right)T_q(\tau)T_r(\xi)T_s(P)\mathrm{w}_t\mathrm{w}_x\mathrm{w}_p dtdxdp=B_tB_xB_p\left(\frac{\pi}{2}\right)^3 S_{qrs}. \quad (10)$$

where

$$\int_{t_0}^{t}F[u]dt'\approx\sum_{k=0}^{K}{}'\sum_{l=0}^{L}{}'\sum_{m=0}^{M}{}' A_{klm}T_k(\tau)T_l(\xi)T_m(P), \quad (11)$$

and

$$\int_{t_0}^{t}sdt'\approx\sum_{k=0}^{K}{}'\sum_{l=0}^{L}{}'\sum_{m=0}^{M}{}' S_{klm}T_k(\tau)T_l(\xi)T_m(P). \quad (12)$$

Combining Eqs. (7a)–(7d), we obtain the final set of algebraic equations

$$a_{qrs}=2\delta_{q0}b_{rs}+A_{qrs}+S_{qrs} \quad (13)$$

where a_{qrs} are the solution coefficients, δ_{q0} is the Kronecker delta function, b_{rs} represents the initial conditions, A_{qrs} is the spectral representation of the time-integrated linear or nonlinear operator F, and S_{qrs} represents the time-integrated source term. Eq. (10) is here defined for the truncated domain $0\leq q\leq K$, whereas strictly the time integration renders $K+1$ terms. For the spatial domain, we have $0\leq r\leq L_{BC}$, where $L_{BC}=L-N_{BC}$, with N_{BC} representing the total number of boundary conditions for this spatial dimension. Boundary condition equations are thus added to (10) for problems including spatial derivatives. Here, we will be solving a set of time-dependent ODEs, thus only initial conditions are required to find a unique solution. Finally, it holds that $0\leq s\leq M$.

Eq. (10) can be solved iteratively with a suitable root solver. Here, we use the semi-implicit root solver (SIR) [11] because it shows superior global stability compared to the Newton method (NM) with line-search while still maintaining a fast convergence.

For more details regarding GWRM basics, please see references [7, 12].

3. The GWRM, exemplified by the Lorenz 1984 equations

The Lorenz 1984 model is a set of three nonlinear ordinary differential equations that features chaotic behavior similar to that of meteorological systems [8]. It is consequently one of the simplest models of Hadley circulation due to the substantial amount of approximations postulated in order to arrive at this low order model. Thus, whereas the Lorenz 1984 model is not an accurate NWP model, it allows for a rigorous numerical analysis of simple, yet nonlinear chaotic behavior. This is the reason for employing them in the present work. The equations are

$$\begin{aligned}\frac{dX}{dt} &= -Y^2 - Z^2 - \alpha X + \alpha Q,\\ \frac{dY}{dt} &= XY - \beta XZ - Y + W,\\ \frac{dZ}{dt} &= \beta XY + XY - Z.\end{aligned} \tag{14}$$

For a more exhaustive description of the model, see [8]. Suffice it to say the variables X, Y, and Z represent certain meteorological systems such as wind currents and large-scale eddies. The coefficients α, β, Q, and W are chosen within certain limits to act as damping, coupling, and amplification of the physical processes of the system.

Lorenz, in his 1984 article, posed the question, "What can such a simple model possibly tell us about the real atmosphere?". His answer was that postulating certain hypothesis about the behavior of the real atmosphere, we can strengthen the reasoning behind the hypothesis qualitatively by the help of these models. However, that is not what we are concerned with here, but rather the model is used as a test to show how different numerical methods handle chaotic behavior. Furthermore, the model is useful to develop new techniques that can analyze such quantities as error growth, parameter uncertainties, and variable perturbations in a more efficient way.

3.1. Time intervals

The GWRM is a global method where a finite number of Chebyshev modes are used to represent the physics in a given temporal domain. Instead of using a large number of Chebyshev modes to obtain a single solution for the entire temporal domain, it is more efficient to split the domain into smaller subintervals, so that lesser Chebyshev modes can be used in each interval.

For controlling and maintaining the same numerical accuracy within each time interval, an automatic *time-adaptive* algorithm is employed. Time-adaptive techniques are easily implemented with Chebyshev polynomials. Since the absolute value of the Chebyshev coefficients decrease with increasing modes for well-resolved functions, the ratio of the absolute values of the highest modes to those of the lowest modes (see Eq. (12)) gives a direct estimate of how "good" the solution is. Thus, it holds that (for the example that (10) represents an ODE)

$$R_K = \frac{|a_K| + |a_{K-1}|}{|a_0| + |a_1|} \to 0, \quad K \to \infty \tag{15}$$

where a_k, $k = 0..K$, are the Chebyshev coefficients with K as the highest mode. By requiring that $R_K < \epsilon$, where ε is a stipulated accuracy, the time subintervals to achieve this accuracy are automatically computed. We have found it efficient to use a balanced algorithm that tries to expand the time interval at about every 10th time interval by a factor 1.5, and that halves the interval whenever ϵ accuracy could not be achieved.

An interesting and useful property of Chebyshev polynomials is $T_k(1) = 1$, so that

$$u(t_1) \approx \sum_{k=0}^{K}{}' a_k T_k(1) = \sum_{k=0}^{K}{}' a_k. \tag{16}$$

Thus, the exact solution at the end state $u(t_1)$ can be approximated by a sum of the Chebyshev coefficients. Eq. (13) conveniently allows for the previous end conditions, represented as Chebyshev coefficients, to be applied as initial conditions for the next interval. All computations of a series of connected intervals are thus calculated in *spectral space*. A GWRM time-interval pseudocode is provided in Appendix A. Only one domain, the temporal, is presented for the sake of clarity.

The pseudocode shows the following: the algebraic equations are collected in a vector ϕ, along with the initial conditions in the first element of the vector. The time domain is divided into subintervals in which the ϕ equations are solved by SIR. The initial condition is equal to the end-state of the previous solution, see (13). If the convergence parameter "*conv*" of the solution is larger than the pre-set minimum error, then the current subinterval is halved and the ϕ equations are solved by SIR. On the other hand, if the "*conv*" convergence criterion is satisfied, the algorithm immediately proceeds to solve the next time interval, the length of which is increased by a factor 1.5 typically every 10th interval. This procedure is done until the entire temporal domain is solved for. When all coefficients are solved for, the semianalytical solution ansatz is easily computed.

3.2. Parameter dependency

Instruments are continually measuring the weather's temperature, wind speeds, and topological height and width to name a few parameters. However, the measurement devices cannot, at present, give a global coverage. Thus, meteorological parameters are approximate and interpolated.

The meteorological parameters are represented in the Lorenz model by α, β, Q, and W. The uncertainty of the measured parameters will inevitably lower the confidence of the solution, specifically important in weather prediction.

When studying parameter dependencies, the GWRM enables an interesting solution. Instead of assigning the parameter a single value, we can introduce a *spectrum of values*. Thus, we introduce a parameter dimension. This technique was first presented in [7], where it was applied to the viscosity parameter in the 1D nonlinear Burger equation. Later, it was applied to a two-dimensional magnetohydrodynamic problem in the paper by Riva et al. [13].

For the Lorenz ODEs, where we have temporal and parameter dependencies, the solution ansatz then takes the form,

$$u(t;\alpha) \approx U(\tau;P) = \sum_{k=0}^{K}{}' \sum_{m=0}^{M}{}' a_{km} T_k(\tau) T_m(P). \tag{17}$$

In Eq. (14), a single parameter has been included; however, the procedure can be expanded to any number of parameters. The impact of the parameter α from the Lorenz equation (11) on the X variable is demonstrated in **Figure 1**. The result from a single GWRM computation is displayed, using the ansatz (14). It can be seen that varying α with the other parameters of Eq. (11) held fixed, the solution is strongly dependent on α for longer times. For lower values of α, the solution becomes more stable and fluctuates with a higher frequency than for higher values of α.

It is of interest to provide a quantitative measure of the effect of α on the solution. One approach would be to compare the deviations from a "base" run for different values of α. For simplicity, however, we have chosen to monitor the standard deviations as a measure of the deflection from the average value in the entire parameter interval. Standard deviations can be computed from the semianalytical solutions with the formula

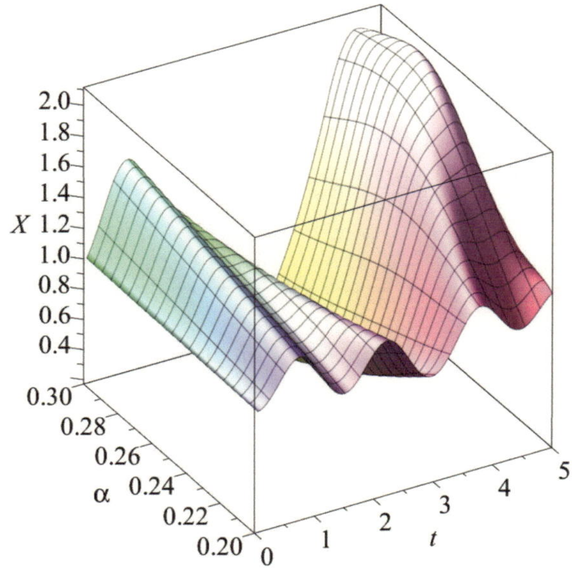

Figure 1. GWRM solution $X(t;\alpha)$ showing the parameter α dependency (interval [0.2, 0.3]) on variable X in the time interval $t \in [0, 5]$. The Lorenz parameters used are $\beta = 4.0$, $Q = 8.0$, and $W = 1.0$. GWRM parameters used are $M = 8$ and $K = 8$.

$$\sigma = \sqrt{\frac{\sum_{i=1}^{N} \left(\mathrm{U}_i - \overline{\mathrm{U}}\right)^2}{N-1}}, \tag{18}$$

where U_i is the GWRM solution for a specific value of the parameter. The evaluations have been carried out at N parameter points i at a time of interest. The parameter interval averaged solution is denoted by $\overline{\mathrm{U}}$. A table is presented below where comparisons are made with a base run with specified parameter values α^*, β^*, Q^*, and W^*. The solution at $t = 1.0$ is used. Average values $\overline{\mathrm{U}}$ and standard deviations σ, representing the entire parameter interval are also given.

The standard deviations are supplied with indices to indicate the chosen parameter and variable for analysis. For example, $\sigma_{\alpha,x}$ states that α is the parameter domain and the standard deviation of X is being analyzed. One parameter at a time has been varied by ±20%, while the others are kept constant.

Parameter dependency example, using a single GWRM computation $t = 1.0$, $K = 8$, $M = 8$

GWRM sol.	$X(1;P^*) = 1.311$	$Y(1;P^*) = -0.046$	$Z(1;P^*) = 1.541$
	$\alpha = [0.2, 0.3]$, $\alpha^* = 0.25$		
$\overline{U} \pm \sigma$	$\overline{X} \pm \sigma_{\alpha,x}$	$\overline{Y} \pm \sigma_{\alpha,y}$	$\overline{Z} \pm \sigma_{\alpha,z}$
	1.307 ± 0.079	-0.061 ± 0.353	1.498 ± 0.073
	$\beta = [3.2, 4.8]$, $\beta^* = 4.0$		
$\overline{U} \pm \sigma$	$\overline{X} \pm \sigma_{\beta,x}$	$\overline{Y} \pm \sigma_{\beta,y}$	$\overline{Z} \pm \sigma_{\beta,z}$
	1.322 ± 0.068	-0.036 ± 0.584	1.401 ± 0.128
	$Q = [6.4, 9.6]$, $Q^* = 8.0$		
$\overline{U} \pm \sigma$	$\overline{X} \pm \sigma_{Q,x}$	$\overline{Y} \pm \sigma_{Q,y}$	$\overline{Z} \pm \sigma_{Q,z}$
	1.308 ± 0.096	-0.071 ± 0.414	1.482 ± 0.087
	$W = [0.8, 1.2]$, $W^* = 1.0$		
$\overline{U} \pm \sigma$	$\overline{X} \pm \sigma_{W,x}$	$\overline{Y} \pm \sigma_{W,y}$	$\overline{Z} \pm \sigma_{W,z}$
	1.310 ± 0.010	-0.044 ± 0.110	1.536 ± 0.015

It is immediately seen that the Y variable is strongly affected by changes in all parameter values, whereas the X and Z solutions are more robust.

3.3. Initial condition uncertainty

Meteorological models generally include time-dependent PDEs and ODEs that require initial conditions as starting points. In mathematical terms, initial conditions are required to close the system of equations. If we are interested in computing a family of scenarios employing different initial conditions, the standard approach to determine the final states is to restart the computations from scratch, using a new initial condition each time.

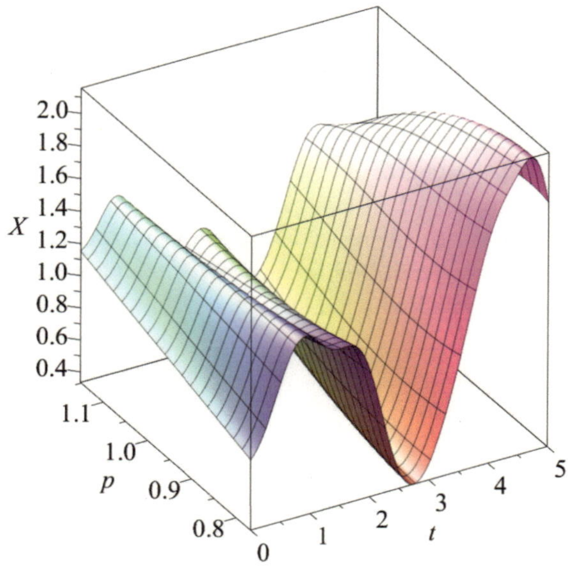

Figure 2. GWRM solution $X(t; P)$ of the Lorenz equations in the time interval $t \in [0, 5]$, where P is the initial condition parameter varying in the interval $[0.768, 1.152]$. The Lorenz parameters used are $\alpha = 0.25$, $\beta = 4.0$, $Q = 8.0$, and $W = 1.0$. GWRM parameters used; $M = 8$ and $K = 8$.

Time-spectral methods offer a more convenient approach. Similarly, as when we computed parameter dependency in a single GWRM computation in the previous section, we will now study the effect of a spectrum of initial values on the solution of Eq. (11), employing a single GWRM computation.

Initial condition dependency example, using a single GWRM computation $t = 1.0, K = 8, M = 8$			
GWRM sol.	$X(1; P^*) = 1.311$	$Y(1; P^*) = -0.046$	$Z(1; P^*) = 1.541$
	$X(0; P) \in [0.768, 1.152]$, $X^* = 0.96$		
$\overline{U} \pm \sigma$	$\overline{X} \pm \sigma_{x,x}$	$\overline{Y} \pm \sigma_{x,y}$	$\overline{Z} \pm \sigma_{x,z}$
	1.306 ± 0.079	-0.047 ± 0.205	1.521 ± 0.087
	$Y(0; P) \in [-1.32, -0.88]$, $Y^* = -1.1$		
$\overline{U} \pm \sigma$	$\overline{X} \pm \sigma_{y,x}$	$\overline{Y} \pm \sigma_{y,y}$	$\overline{Z} \pm \sigma_{y,z}$
	1.319 ± 0.156	-0.021 ± 0.440	1.460 ± 0.092
	$Z(0; P) \in [0.4, 0.6]$, $Z^* = 0.5$		
$\overline{U} \pm \sigma$	$\overline{X} \pm \sigma_{z,x}$	$\overline{Y} \pm \sigma_{z,y}$	$\overline{Z} \pm \sigma_{z,z}$
	1.309 ± 0.008	-0.041 ± 0.132	1.535 ± 0.008

An interesting, albeit challenging, feature of meteorological models is the *inherent uncertainty* of the initial conditions. As a result, how certain can we be of our numerical results? To be able to predict a chaotic system with absolute precision, one would need infinitely accurate measuring devices, computer, and numerical method. Since none of these are granted, error growth analysis is important to gauge how far into the future we can be confident of our predictions.

A classical error growth analysis employed for the Lorenz equations (11) would be the following. A base scenario $U_{k,1}$ (where "k" denotes the variable) with initial conditions $\mathbf{v} =< X(0), Y(0), Z(0) >=< 0.96, -1.10, 0.50 >$ with a time window of $t \in [0, 50]$ is solved for. Subsequently, a number of perturbed scenarios N are solved, denoted with superscript ($'$), where the initial conditions are perturbed an amount δ, taking values < 0.001, for example. The error growth is then computed with the formula,

$$E_k(t) = \frac{1}{N} \sum_{n=2}^{N} \left(U'_{k,n} - U_{k,1} \right)^2, \quad k = 1, 2, \text{and } 3. \tag{19}$$

Thus, in this traditional analysis, a large number of computations are needed, where the ODEs are solved for different perturbed initial conditions.

We suggest another approach. By use of an ansatz of the type (14), a spectrum of initial conditions is allowed in a single computation. In the table above, results are presented for three cases where Eq. (11) has been solved to time $t = 1.0$. For the three cases, the base initial conditions X^*, Y^*, and Z^* as well as intervals for $X(0, P)$, $Y(0, P)$, and $Z(0, P)$ are shown. Also, results for the base initial conditions at $t = 1$ are provided. For the solutions, where the initial

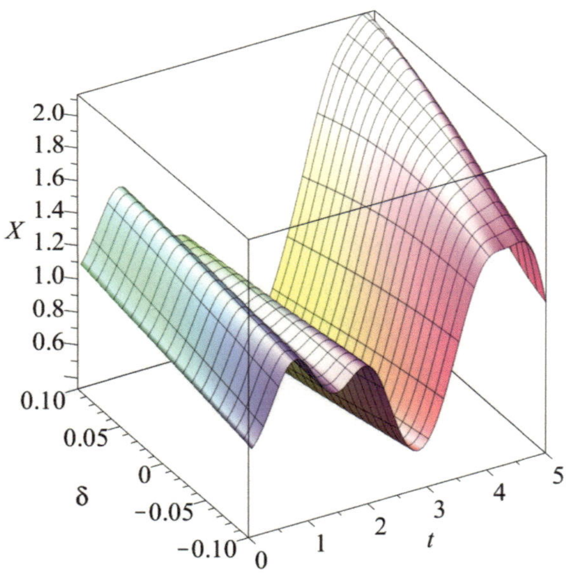

Figure 3. This figure shows the result of simultaneously varying the initial conditions of all the variables X, Y, and Z. The plot is of the GWRM solution $X(t; \delta)$ in the time interval $t \in [0, 5]$, where δ is the perturbation parameter with interval $[-0.1, 0.1]$ applied to the initial condition $X'(0; \delta)$. The Lorenz parameters used are $\alpha = 0.25$, $\beta = 4.0$, $Q = 8.0$, and $W = 1.0$. GWRM parameters used are $M = 8$ and $K = 8$.

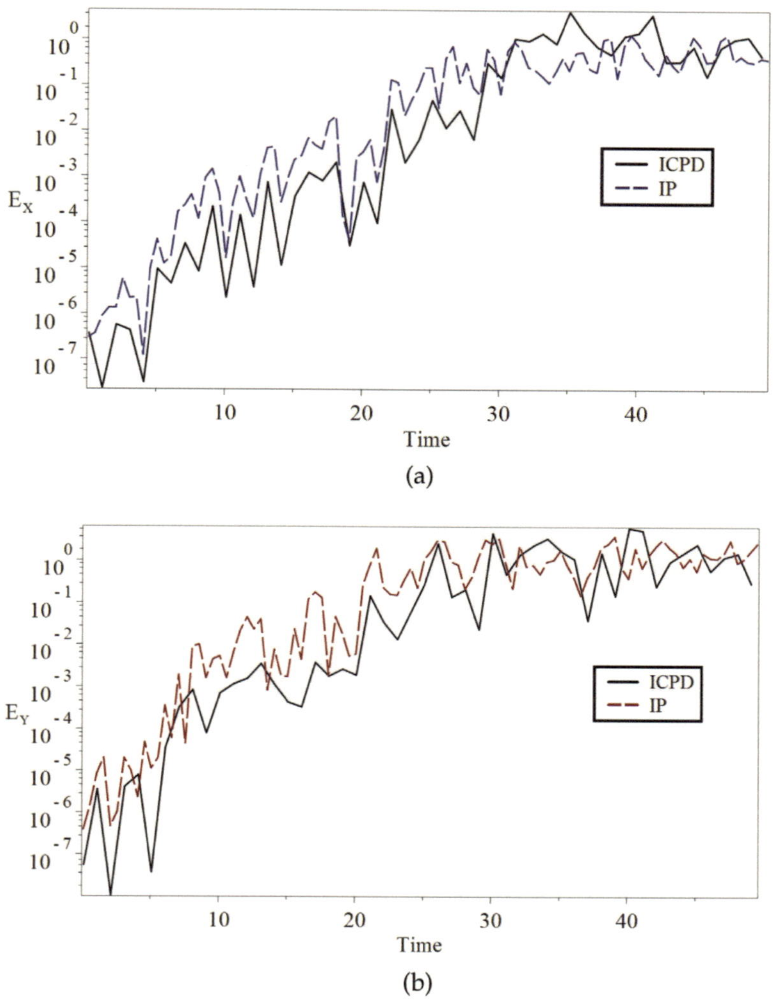

Figure 4. Comparison of error growth for (a) X and (b) Y (16) for the ICPD and PI schemes, applied to the Lorenz equation (11). (Solid) ICPD computation requiring $t_{CPU} = 2.7$ min and (dashed) PI computation requiring $t_{CPU} = 12.7$ min. The Lorenz parameters used are $\alpha = 0.25$, $\beta = 4.0$, $Q = 8.0$, and $W = 1.0$. GWRM parameters used are $K = 8$ and $M = 4$.

conditions are allowed to vary in an interval, both averages over the interval are shown as well as corresponding standard deviations. The analysis shows that the Y variable value at $t = 1$ is strongly dependent on the initial condition. In **Figure 2**, the time and initial condition dependence of the X variable is shown in a 3D diagram. Here, the run time has been extended to $t = 5$.

It would also be of interest to study the effect of allowing the initial conditions for all variables X^*, Y^*, and Z^* to vary simultaneously. In order to accomplish this in a single GWRM run, the natural thing to do would be to extend the ansatz (14) to include three parameters, corresponding to the different initial conditions. Instead, we have chosen a simpler approach.

We let $X'(0;\delta) = X^* + \delta$, $Y'(0;\delta) = Y^* + \delta$, and $Z'(0;\delta) = Z^* + \delta$, where the perturbation δ is a single parameter that is applied to all three variables. An example computation is shown in **Figure 3**. The X variable here features a quite a different time evolution than in **Figure 2**, where only the X variable initial condition was perturbed.

In the table below, initial condition interval average values, as well as standard deviations, are provided for all the variables X, Y, and Z for a run extending to $t = 20$.

Initial condition dependency example, using a single GWRM computation $t = 20$, $K = 8$, $M = 8$, $\delta < 0.001$			
GWRM sol.	$X(20;0) = 1.779$	$Y(20;0) = 0.483$	$Z(20;0) = -0.167$
$\overline{U} \pm \sigma$	$\overline{X} \pm \sigma_x$	$\overline{Y} \pm \sigma_y$	$\overline{Z} \pm \sigma_z$
	1.787 ± 0.063	0.460 ± 0.062	-0.099 ± 0.226

Returning to error growth analysis as suggested by Eq. (16), reliable results can be computed if a sufficiently large domain has been spanned by the perturbed initial conditions. We have compared the classical approach of perturbed iterations (PI) with the new initial condition parameter dependency (ICPD) technique in **Figure 4a** and **b**. Of particular interest is the potential gain in CPU time, using the ICPD.

It is found that the ICPD technique with $K = 8$ and $M = 4$ achieves the same result of error growth as the PI scheme. The ICPD scheme computed the error growth in $t_{CPU} = 2.7$ min as compared to the PI scheme which required $t_{CPU} = 12.7$ min. Thus, a near fivefold increase in efficiency was obtained for this relatively simple case.

Numerically, **Figure 4a** is based on $N = 250$ runs with the PI scheme, whereas **Figure 4b** was computed in a single run, where $N = 1000$ different initial condition values was used.

4. Discussion

Is it more *efficient* to solve a system of differential equations using a spectral representation of the parameter domain instead of solving the system multiple times with different parameter values? Some points related to this discussion are that for the ICPD approach, using the GWRM,

- the parameter dependencies can easily be analyzed since GWRM solutions are analytical
- the time-spectral method is a high-order method, leading to high accuracy solutions
- all parameters in an interval are included, whereas certain regions of critical parameter dependence could be missed out with traditional PI methods
- the parameter domain can be split into intervals, that can be solved for separately, potentially spanning a larger parameter space more efficiently

In this chapter, a univariate analysis has been performed; however, a multivariate analysis can also be implemented in the same manor. It should be noted, however, that the more parameters introduced in the analysis, the more Chebyshev modes need to be solved for, which results in larger matrices and memory demands.

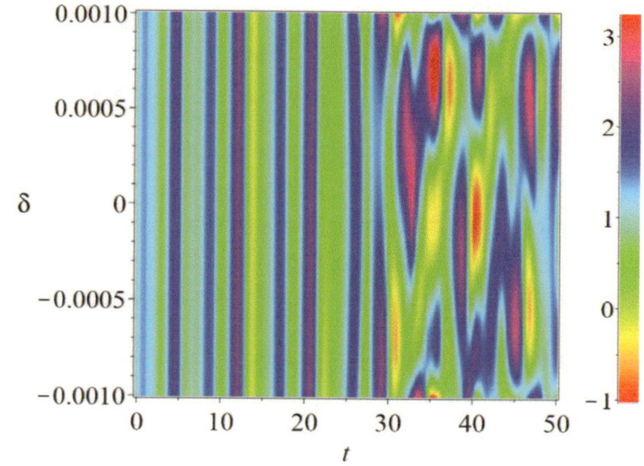

Figure 5. Density plot of $X(t;\delta)$ in the time interval $t \in [0, 50]$, where δ is perturbation parameter with interval $[-0.001, 0.001]$ applied to the initial condition $X'(0;\delta)$. GWRM parameters used; $K = 8$ and $M = 4$.

It is also of interest to address the *accuracy* of the present computations. The ICPD with $K = 8$ and $M = 4$ did not achieve high accuracy of the exact solution at the end time $t = 50$ as can be seen in **Figure 5**. To be more precise, the solution at higher times ($\sim t > 25$) does not represent the real dynamics because the introduced error has grown to an extent where the real solution is lost (see **Figure 5**). For long run times, the effect of uncertain initial conditions is muddled by the effect of numerical inaccuracy. On top of this, the introduction of the parameter dimension itself in the ansatz (14) is a source for inaccuracy, since the GWRM algorithm has to handle a larger set of Chebyshev polynomials in this case. The argument could be made that if a lower order (e.g., fourth order) solution ansatz cannot accurately represent the parameter "physics," then the solution exhibits no predictive power if the *prediction horizon* has been exceeded. How could then the error growth be more accurately represented? The parameter Chebyshev coefficient could perhaps be increased, or the parameter domain could be split into intervals. These are interesting questions for future studies of ICPD methods related to NWP modeling.

5. Conclusion

Spectral methods have a long history when applied to the spatial domain in PDE modeling. The time-spectral method GWRM, demonstrated here, provides similar accuracy for the temporal domain. High-order methods, like the GWRM, may achieve resolutions much higher than that of finite difference methods for similar amount of work. Furthermore, when applied to meteorological systems, the time-spectral method can accurately and efficiently compute the physics in all spatial, temporal, and parameter domains.

A recent study, comparing the time-spectral method GWRM with commonly used time-stepping methods such as Runge-Kutta and other, high-order implicit methods, was carried out in [1]. Since then, the GWRM numerical algorithms have been further streamlined; see for example [14]. Moreover, it has been found that the Jacobian of the algebraic system of equations (10) need only be computed once by including the time interval length analytically. Thus, during the computations, the new time interval length can be substituted into the Jacobian before use. This decreases the computation time from $t_{CPU} = 17.9$ min to $t_{CPU} = 12.7$ min, roughly 30%, for 250 perturbed scenarios with $K = 8$ and $t \in [0, 50]$.

We have, in this chapter, also presented a new approach to error analysis using the initial condition parameter dependency (ICPD) technique, that is, by spectral representation of the parameter domain. In a single computation, the same error growth was reproduced as for the classical case (where a large number of runs were carried out for different initial conditions), with a computational time of $t_{CPU} = 2.7$ min. This amounts to a near fivefold gain in CPU time efficiency.

Finally, it was shown that the ICPD technique is also successfully applicable for determining the effect of perturbed physical variables on the solution. In one GWRM computation only, it was found that the solutions to the Lorenz 1984 equations sensitively depend on the system parameter α.

A. Pseudo-code

Algorithm 1. GWRM time Intervals.

```
procedure
    φ ← A_k        Algebraic equations
    φ[0] ← A_0 + 2.0 · IC   Apply initial condition IC
    t_acc ← 0
    j ← 1
    if t_acc > Time them
        conv ← 1.
        while conv > rel.error do
            if j > 1 them
                φ[0] ← A_0 + 2.0(x_{j-1}[0]/2 + x_{j-1}[1] + ... + x_{j-1}[K].
            x_j ← SIR(φ.
            conv ← (|x[K]| + |x[K − 1]|)/(|x[0]| + |x[1]|).
            if conv < rel.error them
                a_k ← x_k        Save solution
            else
                Δt ← Δt/2
        t_acc ← t_acc + Δt
        if modp(j, 10) = 0 them
            Δt ← 1.5Δt
        j ← j + 1
```

http://dx.doi.org/10.5772/intechopen.80351

Author details

Kristoffer Lindvall* and Jan Scheffel

*Address all correspondence to: kristoffer.lindvall@ee.kth.se

KTH Royal Institute of Technology, Sweden

References

[1] Scheffel J, Lindvall K, Yik HF. A time-spectral approach to numerical weather prediction. Computer Physics Communications. 2018;**226**:127-135

[2] Morchoisne Y. Resolution of the Navier-Stokes equations by a space-time pseudospectral method. La Recherche Aerospatiale. 1979;**5**:293-306

[3] Peyret R, Taylor TD. Computational Methods for Fluid Flow. New York: Springer; 1983

[4] Tal-Ezer H. Spectral methods in time for hyperbolic equations. SIAM Journal on Numerical Analysis. 1986;**23**:1-26

[5] Tal-Ezer H. Spectral methods in time for parabolic problems. SIAM Journal on Numerical Analysis. 1989;**26**:1-11

[6] Bar-Yoseph P, Moses E, Zrahia U, Yarin A. Space-time spectral element methods for one-dimensional nonlinear advection-diffusion problems. Journal of Computational Physics. 1995;**119**:62

[7] Scheffel J. Time-spectral solution of initial-value problems. In: Jang C-L, editor. Partial Differential Equations: Theory, Analysis and Applications. New York: Nova Science Publishers, Inc.; 2011. pp. 1-49

[8] Lorenz EN. Irregularity: A fundamental property of the atmosphere. Tellus A: Published online 2016. 1984;**36**:98-110

[9] Mason JC, Handscomb DC. Chebyshev Polynomials. London: Chapman and Hall; 2003

[10] Forsyth AR. Calculus of Variations. New York: Dover Publications Inc.; 1960

[11] Scheffel J, Lindvall K. SIR—An efficient solver for systems of equations. SoftwareX. 2018;**7**: 59-62

[12] Scheffel J. A spectral method in time for initial-value problems. American Journal of Computational Mathematics AJCM. 2012;**2**:173-193

[13] Riva F, Milanese L, Ricci P. Uncertainty propagation by using spectral methods: A practical application to a two-dimensional turbulence fluid model. Physics of Plasmas. 2017;**24**:1-12

[14] Scheffel J, Lindvall K. J. Scheffel and K. Lindvall. Optimizing Time-Spectral Solution of Initial-Value Problems. American Journal of Computational Mathematics AJCM. 2018; **8**:7-26

Chapter 6

Atmospheric Radiative Transfer Parameterizations

Feng Zhang, Yi-Ning Shi, Kun Wu, Jiangnan Li and
Wenwen Li

Additional information is available at the end of the chapter

http://dx.doi.org/10.5772/intechopen.82122

Abstract

Various radiative transfer (RT) schemes are presented in the chapter including four-stream discrete ordinates adding method (4DDA), four-stream harmonic expansion approximation (4SDA) for the solar spectra and absorption approximation (AA), variational iteration method (VIM) for the infrared spectra. 4DDA uses Gaussian quadrature method to deal with the integration in the RT equation. 4SDA considers four-order spherical harmonic expansion in radiative intensity. VIM allows the zeroth-order solution to be identified as AA, and the scattering effect is included in the first-order iteration. By applying 4DDA/4SDA to a realistic atmospheric profile with gaseous transmission considered, it is found that the accuracy of 4DDA/4SDA is superior to two stream spherical harmonic (Eddington approximation) adding method (2SDA) and two-stream discrete ordinates adding method (2DDA), especially for the cloudy conditions. It is shown that the relative errors of 4DDA/4SDA are generally less than 1% in heating rate, while the relative errors of both 2SDA and 2DDA are over 6%. By applying VIM to a full radiation algorithm a gaseous gaseous transmission, it is found that VIM is generally more accurate than the discrete ordinates method (DOM). Computationally, VIM is slightly faster than DOM in the two-stream case but more than twice as fast in the four-stream case. In view of its high overall accuracy and computational efficiency, 4DDA, 4SDA, as well as VIM are well suited in solving radiative transfer in climate models.

Keywords: atmospheric radiative transfer, solar four-stream discrete ordinates adding method, solar four-stream harmonic expansion approximation, infrared variational iteration method, infrared absorption approximation, integro-differential equation

1. Introduction

Solving the radiative transfer (RT) equation is a key issue when dealing with solar/infrared radiative processes related to climate simulations (e.g., radiative flux and heating rate calculation at each level of a climate model). In recent decades, numerous approximation techniques have been proposed to solve the RT equation [1–5].

For solar radiation, analytical solutions of various two-stream approximations [6] have been widely used in climate models [7]. More accurate methods of four-stream approximation and six-stream approximation have also been proposed (e.g., [8–13]). Based on the invariance principle [14], an adding method of four-stream discrete ordinates method (DOM) (4DDA) has been proposed. The advantage of the adding method is that it can handle partial cloud in climate models (e.g., [15–17]). 4DDA is much more accurate than the adding method of two-stream DOM (2DDA) in flux and heating rate calculation, especially under cloudy-sky conditions, where cloud plays an important role in radiative transfer with different dynamics and microphysics [18, 19]. Zhang and Li [20] proposed a new solar RT parameterization (4SDA) which is based on four-stream harmonic expansion to simulate RT in climate modeling.

Since scattering is much weaker for infrared radiation than for solar radiation, an absorption approximation (AA) is used in most current climate models [7]. In AA, the scattering phase function is replaced by a δ-function, meaning that scattering is neglected in all but the forward direction. The advantage of AA is that the upward and downward transfer processes are completely separated. If scattering is considered, the upward and downward intensities are coupled. Under this condition, Fu [21] derived an inverse matrix method to deal with multiple scattering in the atmosphere by using the two-stream DOM (2DOM) and the four-stream DOM (4DOM). It is found that 4DOM is more accurate than 2DOM, especially for thin optical depths. Even if scattering is included, TOA errors for 2DOM can still be >2 W m^{-2} for cirrus clouds [22]. The 4DOM results tend to be very accurate, but the calculation process is complicated. Zhang et al. [23] apply the four-stream adding method to resolve multilayer infrared RT.

In the past two decades, the variational iteration method (VIM) has been used to deal with many nonlinear problems [24–31]. In addition, VIM has been shown to converge faster to the exact solution than other methods do. An accurate solution can be found for most nonlinear problems in only one or two iterations for small solution domains in VIM [27, 31]. Because of its flexibility, convenience and efficiency, VIM has been applied to various nonlinear equations ([27–29] and many others). In VIM, a nonlinear problem is separated into linear and nonlinear terms, where the latter are usually difficult to deal with and are initially approximated as a first guess. Subsequently, a correction function is constructed by using a general Lagrange multiplier that can be identified optimally via variational theory. Although the infrared RT equation is not a nonlinear equation, it contains an integral term for the scattering, and this is very complicated to deal with. Therefore, VIM is applied to solve the infrared RT equation.

The objective of the chapter is to introduce 4DDA/4SDA scheme [15] for solar RT, AA, and VIM [32] schemes and apply them to resolve solar/infrared radiative transfer in RT models. The simulation results and their comparison to results from the 128-stream discrete ordinates method radiative transfer scheme [33] are shown in Section 4. Finally, a summary is given in Section 5.

2. Parameterizations for solar radiative transfer

The azimuthally averaged solar RT equation is

$$\mu \frac{dI(\tau,\mu)}{d\tau} = I(\tau,\mu) - \frac{\omega}{2}\int_{-1}^{1} I\left(\tau,\mu'\right)P\left(\mu,\mu'\right)d\mu' - \frac{\omega}{4\pi}F_0 e^{-\tau/\mu_0}P(\mu,-\mu_0) \quad (1)$$

where μ is the cosine of the zenith angle (we use the positive and negative μ to denote the upward and downward light beams, respectively), τ is optical depth, ω is single-scattering albedo, and $P(\mu,\mu')$ is the azimuthally averaged scattering phase function, defining the light incidence at μ', which is scattered in the direction μ. F_0 is the incoming solar flux and μ_0 is the solar zenith angle. The $P(\mu,\mu')$ can be written as

$$P\left(\mu,\mu'\right) = \sum_{l=0}^{N} \omega_l P_l(\mu) P_l\left(\mu'\right) \quad (2)$$

where P_l is the Legendre function. The ω_l can be determined from the orthogonal property of Legendre polynomials: $\omega_l = [(2l+1)/2]\int P(\cos\Theta)P_l(\cos\Theta)d\cos\Theta$. In our notations, $\omega_0 = 1$ and $\omega_1 = 3\text{g}$. g is the asymmetry factor.

2.1. Four-stream discrete ordinates scheme

The four-stream discrete ordinates scheme (4DDA) use the Gaussian quadrature method to deal with the integration in Eq. (1). Eq. (1) yields

$$\mu_i \frac{dI(\tau,\mu_i)}{d\tau} = I(\tau,\mu_i) - \frac{\omega}{2}\sum_{l=0}^{3}\omega_l P_l(\mu_i)\sum_{j=-2}^{2} I\left(\tau,\mu_j\right)P_l\left(\mu_j\right)a_j - \frac{\omega}{4\pi}F_0 e^{-\tau/\mu_0}\sum_{l=0}^{3}\omega_l P_l(\mu_i)P_l(-\mu_0) \quad (3)$$

where the quadrature point $\mu_{-i} = -\mu_i$, the weight $a_{-i} = a_i$ with $i = \pm 1, \pm 2$, $\mu_1 = 0.2113248$, $\mu_2 = 0.7886752$ and $a_1 = a_2 = 0.5$.

After a lengthy and laborious derivation, the solution of (3) is given by

$$\begin{bmatrix} I(\tau,\mu_2) \\ I(\tau,\mu_1) \\ I(\tau,\mu_{-1}) \\ I(\tau,\mu_{-2}) \end{bmatrix} = \begin{bmatrix} \varphi_2^+ e_2 & \varphi_1^+ e_1 & \varphi_1^- e_3 & \varphi_2^- e_4 \\ \varphi_2^+ e_2 & \varphi_1^+ e_1 & \varphi_1^- e_3 & \varphi_2^- e_4 \\ \varphi_2^- e_2 & \varphi_1^- e_1 & \varphi_1^+ e_3 & \varphi_2^+ e_4 \\ \varphi_2^- e_2 & \varphi_1^- e_1 & \varphi_1^+ e_3 & \varphi_2^+ e_4 \end{bmatrix} G + \begin{bmatrix} Z_2^+ \\ Z_1^+ \\ Z_1^- \\ Z_2^- \end{bmatrix} \exp\left(-\tau/\mu_0\right) \quad (4)$$

where G is determined by the boundary conditions, and values of other parameters are shown in [13].

The adding method which is used to deal with multilayer RT is shown in [13].

2.2. Four-stream spherical harmonic expansion scheme

The purpose of the four-stream spherical harmonic expansion (4SDA) is to separate out the angle-dependent intensity by assuming

$$I(\tau,\mu) = \sum_{l=0}^{3} I_l(\tau)P_l(\mu) \tag{5}$$

By substituting Eqs. (2) and (5) into Eq. (1) and using the orthogonality relation of the Legendre function in $-1 \le \mu \le 1$, we obtain

$$\frac{dI_1}{d\tau} = a_0 I_0 - b_0 e^{-\tau/\mu_0} \tag{6}$$

$$2\frac{dI_2}{d\tau} + \frac{dI_0}{d\tau} = a_1 I_1 - b_1 e^{-\tau/\mu_0} \tag{7}$$

$$3\frac{dI_3}{d\tau} + 2\frac{dI_1}{d\tau} = a_2 I_2 - b_2 e^{-\tau/\mu_0} \tag{8}$$

$$3\frac{dI_2}{d\tau} = a_3 I_3 - b_3 e^{-\tau/\mu_0} \tag{9}$$

The solution of Eqs. (6)–(9) is

$$\begin{bmatrix} I_0 \\ I_1 \\ I_2 \\ I_3 \end{bmatrix} = \begin{bmatrix} e_1 & e_3 & e_2 & e_4 \\ R_1 e_1 & -R_1 e_3 & R_2 e_2 & -R_2 e_4 \\ \hat{P}_1 e_1 & \hat{P}_1 e_3 & \hat{P}_2 e_2 & \hat{P}_2 e_4 \\ Q_1 e_1 & -Q_1 e_3 & Q_2 e_2 & -Q_2 e_4 \end{bmatrix} G + \begin{bmatrix} \eta_0 \\ \eta_1 \\ \eta_2 \\ \eta_3 \end{bmatrix} \exp\left(-\tau/\mu_0\right) \tag{10}$$

where G is determined by the boundary conditions, and values of other parameters can be found at [20].

The 4SDA can also be applied to solve multilayer solar RT. The detailed process can be found at [20].

3. Parameterizations for infrared radiative transfer

The infrared RT equation for intensity $I(\tau,\mu)$ is

$$\mu\frac{dI(\tau,\mu)}{d\tau} = I(\tau,\mu) - (1-\omega)B(\tau) - \frac{\omega}{2}\int_{-1}^{1} I\left(\tau,\mu'\right)P\left(\mu,\mu'\right)d\mu' \tag{11}$$

where μ, τ, ω, and P are the same as one in Eq. (1). The Planck function is approximated exponentially in optical depth [17, 18] as $B(\tau) = B_0 e^{-\beta\tau}$ where $\beta = -[\ln(B_1/B_0)]/\tau_1$. τ_1 is the optical depth of the considered layer. Planck functions B_0 and B_1 are evaluated by the temperature of the top ($\tau = 0$) and bottom ($\tau = \tau_1$) of the layer, respectively.

3.1. Absorption approximation

In absorption approximation (AA), the scattering phase function $P(\mu, \mu')$ is simplified as a $\delta(\mu, \mu')$ function [22], and the infrared RT equation becomes

$$\mu \frac{dI^0(\tau, \mu)}{d\tau} = (1-\omega) I^0(\tau, \mu) - (1-\omega) B_0 e^{-\beta\tau} \tag{12}$$

We use I^0 as the radiative intensity from the absorption approximation. The solution of Eq. (12) is

$$I^0(\tau, \pm\mu) = \left[I^0(\tau_s, \pm\mu) \mp \frac{1-\omega}{\mu\beta \pm (1-\omega)} B_0 e^{-\beta\tau_s} \right] e^{\pm(1-\omega)(\tau-\tau_s)/\mu} \pm \frac{1-\omega}{\mu\beta \pm (1-\omega)} B_0 e^{-\beta\tau} \tag{13}$$

where $\pm\mu$ is corresponding to the upward (downward) path and τ_s is the initial point at optical depth $\tau_s = 0$ for the downward direction and at $\tau_s = \tau_1$ for the upward direction. Generally, AA is applied to the two-stream case (δ-2AA) and the four-stream case (δ-4AA). In the following subsection, the AA solution is used as the initial approximate solution of the VIM.

3.2. Variational iteration method

A general nonlinear system is used to illustrate the basic idea of variational iteration method (VIM) [24–31]:

$$Lv(x) + Nv(x) = h(x) \tag{14}$$

where $v(x)$ is the function to be solved and $h(x)$, L, and N are an inhomogeneous term, a linear operator, and a nonlinear operator, respectively. If $v(x)$ is found in the (n)th iteration, the (n + 1)th-order functional solution is

$$v^{n+1}(x) = v^n(x) + \int_0^x \lambda(\zeta)[Lv^n(\zeta) + N\tilde{v}^n(\zeta) - h(\zeta)]d\zeta \tag{15}$$

Here, $\tilde{v}^n(\zeta)$ is considered as a restricted variation [i.e., $\hat{\delta}\tilde{v}^n(\zeta) = 0$], where $\hat{\delta}$ represents the variation. The term $\lambda(\zeta)$ is a general Lagrange multiplier and should be identified optimally by using variational theory.

Based on VIM, the functional reiteration of the infrared RT equation can be deduced as

$$I^{n+1}(\tau, \pm\mu) = I^n(\tau, \pm\mu) + \int_{\tau_s}^{\tau} \lambda^{\pm}(s) \left[\frac{dI^n(s, \pm\mu)}{ds} \mp \frac{I^n(s, \pm\mu)}{\mu} \pm \frac{\omega}{2\mu} \int_{-1}^{1} \tilde{I}n(s, \mu') P(\pm\mu, \mu') d\mu' \right.$$
$$\left. \pm (1-\omega) B_0 e^{-\beta s/\mu} \right] ds \tag{16}$$

According to VIM theory, under the conditions that $\hat{\delta} I^n(s, \pm\mu)\Big|_{s=\tau_s} = 0$, $\hat{\delta} \int_{\tau_s}^{\tau} \frac{(1-\omega)B_0 e^{-\beta s}}{\mu} ds = 0$ and the restricted variation $\hat{\delta}\tilde{I}^n(s, \mu') = 0$, Eq. (16) yields

$$\hat{\delta} I^{n+1}(\tau, \pm\mu) = \left[1 + \lambda^{\pm}(s)\right]\hat{\delta} I^n(s, \pm\mu)\Big|_{s=\tau} + \hat{\delta}\int_{\tau_s}^{\tau}\left[-\frac{d\lambda^{\pm}(s)}{ds} \mp \frac{\lambda^{\pm}(s)}{\mu}\right] I^n(s, \pm\mu) ds \quad (17)$$

For the above correction functional to be stationary [i.e., $\hat{\delta} I^{n+1}(\tau, \pm\mu) = 0$], we require the following stationary conditions:

$$-\frac{d\lambda^{\pm}(s)}{ds} \mp \frac{\lambda^{\pm}(s)}{\mu} = 0 \text{ and } \lambda^{\pm}(s)\Big|_{s=\tau} = -1 \quad (18)$$

Therefore the Lagrangian multiplier $\lambda^{\pm}(s)$ can be identified as

$$\lambda^{\pm}(s) = -e^{\pm(\tau - s)/\mu} \quad (19)$$

By substituting Eq. (19) into Eq. (16), we obtain

$$I^{n+1}(\tau, \pm\mu) = I^n(\tau, \pm\mu) - \int_{\tau_s}^{\tau} e^{\pm(\tau-s)/\mu}\left[\frac{dI^n(s, \pm\mu)}{ds} \mp \frac{I^n(s, \pm\mu)}{\mu} \pm \frac{\omega}{2\mu}\int_{-1}^{1} I^n(s, \mu')P(\pm\mu, \mu')d\mu' \pm (1-\omega)B_0 e^{-\beta s/\mu}\right] ds \quad (20)$$

In two-stream VIM (2VIM), the term of $\int_{-1}^{1} I^n(s, \mu')P(\pm\mu, \mu')d\mu'$ in Eq. (22) is simplified as

$$\int_{-1}^{1} I^n(s, \mu')P(\pm\mu, \mu')d\mu' = (1 + 3g\mu\mu_1)I^n(s, \mu_1) + (1 + 3g\mu\mu_{-1})I^n(s, \mu_{-1}) \quad (21)$$

where $\mu_1 = -\mu_{-1} = 1/1.66$. The AA expression shown in Eq. (15) is used as the initial zeroth-order solution. By substituting Eq. (15) into Eq. (22), we obtain the first-order solution of upward intensity at $\tau = 0$ and downward intensity at $\tau = \tau_1$. Detailed process can be found at [30]. The upward and downward fluxes are written as

$$F^{+}(0) = \pi I^1(0, \mu_1) \text{ and } F^{-}(\tau_1) = \pi I^1(\tau_1, \mu_1) \quad (22)$$

We consider a surface boundary condition with a surface emissivity ε_s as follows in 2VIM

$$I_s(\mu_1) = (1 - \varepsilon_s)I_s(\mu_{-1}) + \varepsilon_s B(T_s) \quad (23)$$

where T_s, $I_s(\mu_1)$, and $I_s(\mu_1)$ are the surface temperature, upward intensity, and downward intensity, respectively, at the surface.

In four-stream VIM (4VIM), the term of $\int_{-1}^{1} I^n(s,\mu')P(\pm\mu,\mu')d\mu'$ in Eq. (22) is simplified as

$$\int_{-1}^{1} I^n\left(s,\mu'\right)P\left(\mu_i,\mu'\right)d\mu' = \frac{1}{2}\sum_{j=-2}^{2}\left(1+3g\mu_i\mu_j\right)I^n\left(s,\mu_j\right) \tag{24}$$

where $\mu_1 = -\mu_{-1} = 0.2113248$ and $\mu_2 = -\mu_{-2} = 0.7886752$. The AA expression shown in Eq. (15) is used as the initial zeroth-order solution. Detailed process can be found at [30]. The upward and downward fluxes are written as

$$F^{+}(0) = 2\pi\left[a_1\mu_1 I^1(0,\mu_1) + a_2\mu_2 I^1(0,\mu_2)\right] \tag{25}$$

$$F^{-}(\tau_1) = 2\pi\left[a_1\mu_1 I^1(\tau_1,\mu_{-1}) + a_2\mu_2 I^1(\tau_1,\mu_{-2})\right] \tag{26}$$

The surface boundary condition with an emissivity ε_s is given by

$$I_s\left(\mu_i\right) = 2(1-\varepsilon_s)\sum_{j=1}^{2} a_j\mu_j I_s\left(-\mu_j\right) + \varepsilon_s B(T_s) \tag{27}$$

To enhance the accuracy of the radiative schemes, a δ-function adjustment is used to adjust the optical parameters following Wiscombe [34]. We refer to the VIM solutions with the δ-function applied as δ-2VIM and δ-4VIM.

4. Results and discussion

4.1. Solar spectra

RT in the atmosphere is a complicated process. It depends not only on the single-layer direct reflection and transmission but also on the diffuse results and the gaseous transmission, cloud-aerosol scattering and absorption, etc. It is important to evaluate errors in radiation under a variety of atmospheric conditions by using a radiation algorithm. The Fu-Liou radiation model [35] is used in this study. This model adopts the correlated-k distribution method for gaseous transmission, including five major greenhouse gases, H_2O, CO_2, O_3, NO_2, and CH_4.

In benchmark calculations, the discrete ordinates model [33] with a 128-stream scheme (128S) is incorporated with the gaseous transmission scheme of the Fu-Loiu radiation model [35] by replacing the existing radiative transfer algorithm in the model. Also the two-stream discrete ordinates adding method (2DDA), Eddington adding method (2SDA), four-stream discrete ordinates adding method (4DDA), and four stream spherical harmonic adding method (4SDA) schemes are incorporated with the same gaseous transmission scheme. The atmosphere was vertically divided into 280 layers, each of thickness 0.25 km. The mid-latitude winter of atmospheric profile [36] is used with a surface albedo of 0.2 for each band. For a water cloud the optical properties are parameterized in terms of liquid water content (LWC) and effective radius (re) [36]. Two calculations are performed: (1) a low cloud (LWC = 0.22gm^{-3}, $r_e = 5.89\mu m$) positioned from 1.0 to 2.0 km and (2) a middle cloud (LWC = 0.28gm^{-3}, $r_e = 6.19\mu m$) positioned from 4.0 to 5.0 km. Three solar zenith angles of $\mu_0 = 1$, $\mu_0 = 0.5$, and $\mu_0 = 0.25$ are generally considered.

In **Figure 1**, the benchmark results of heating rate are shown for three solar zenith angles under low/middle cloud condition (**Figure 1a–c** for low cloud condition, **Figure 1g–i** for middle cloud condition) as well as the absolute errors of 2DDA, 2SDA, 4DDA, and 4SDA against the benchmark results (**Figure 1d–f** for cloud condition, **Figure 1j–l** for middle cloud condition). When $\mu_0 = 1$, the absolute errors of 2DDA are up to about 1.5 and 2.3 K day^{-1} (relative errors about 6%) for the low and middle clouds. The error of 2SDA is smaller than 2DDA. When $\mu_0 = 0.25$, the result becomes opposite as the error of 2SDA becomes larger than 2DDA. By using 4DDA and 4SDA, relative errors are much suppressed. In general, the relative error becomes less than 1% for the three solar zenith angles. This indicates both 4DDA and 4SDA are accurate enough to obtain cloud-top solar heating.

4.2. Infrared spectra

For the infrared spectra, the accuracy and efficiency of δ-2AA, δ-4AA, δ-2VIM, δ-4VIM, δ-2DOM, and δ-4DOM will be systematically compared. In addition, the discrete ordinates model [28] with a 128-stream scheme (δ-128DOM) is used as the benchmark model.

A radiation model [17] is used to study the accuracy of the VIM scheme for multiple layers within a model atmosphere. A correlation-k distribution scheme is used to simulate the gaseous transmission with profiles for H_2O, CH_4, CO_2, NO_2, O_3, and CFCs. This model is reasonably efficient because it neglects to scatter for certain intervals with very large absorption coefficients and water vapor continuum at high altitudes. Nine infrared bands are adopted in this model in wavenumber ranges 0–340, 340–540, 540–800, 800–980, 980–1100, 1100–1400, 1400–1900, 1900–2200, and 2200–2500 cm^{-1}. The optical properties of ice and water clouds are calculated based on the radiative property parameterization of [37–39], respectively. The mid-latitude winter atmospheric profiles [36] are used. The atmospheric profile is divided into homogeneous layers with a geometrical thickness of 0.25 km.

In this model, a low cloud with an effective radius $r_e = 5.89\mu m$ and liquid water content LWC = 0.22gm^{-3} is located at 1.0–2.0 km, a middle cloud with $r_e = 5.89\mu m$ and LWC = 0.22 gm^{-3} is located at 4.0–5.0 km, and a high cloud with a mean effective size $D_e = 41.1$ $\mu m$and an ice water content IWC = 0.0048 gm^{-3} is located at 9.0–11.0 km. The surface emissivity ε_s is set to 1.

In the left column of **Figure 2**, the benchmark heating rates calculated by δ-128DOM are given under conditions of low clouds (**Figure 2a**), middle clouds (**Figure 2d**), high clouds (**Figure 2g**) and the all-sky case containing a combination of low, middle, and high clouds (**Figure 2j**). The middle column of **Figure 2** shows the errors in the calculated heating rates of δ-2AA, δ-2DOM, and δ-2VIM compared to those of δ-128DOM. Furthermore, the right column of **Figure 2** shows the errors in the calculated heating rates of δ-4AA, δ-4DOM, and δ-4VIM compared to those of δ-128DOM.

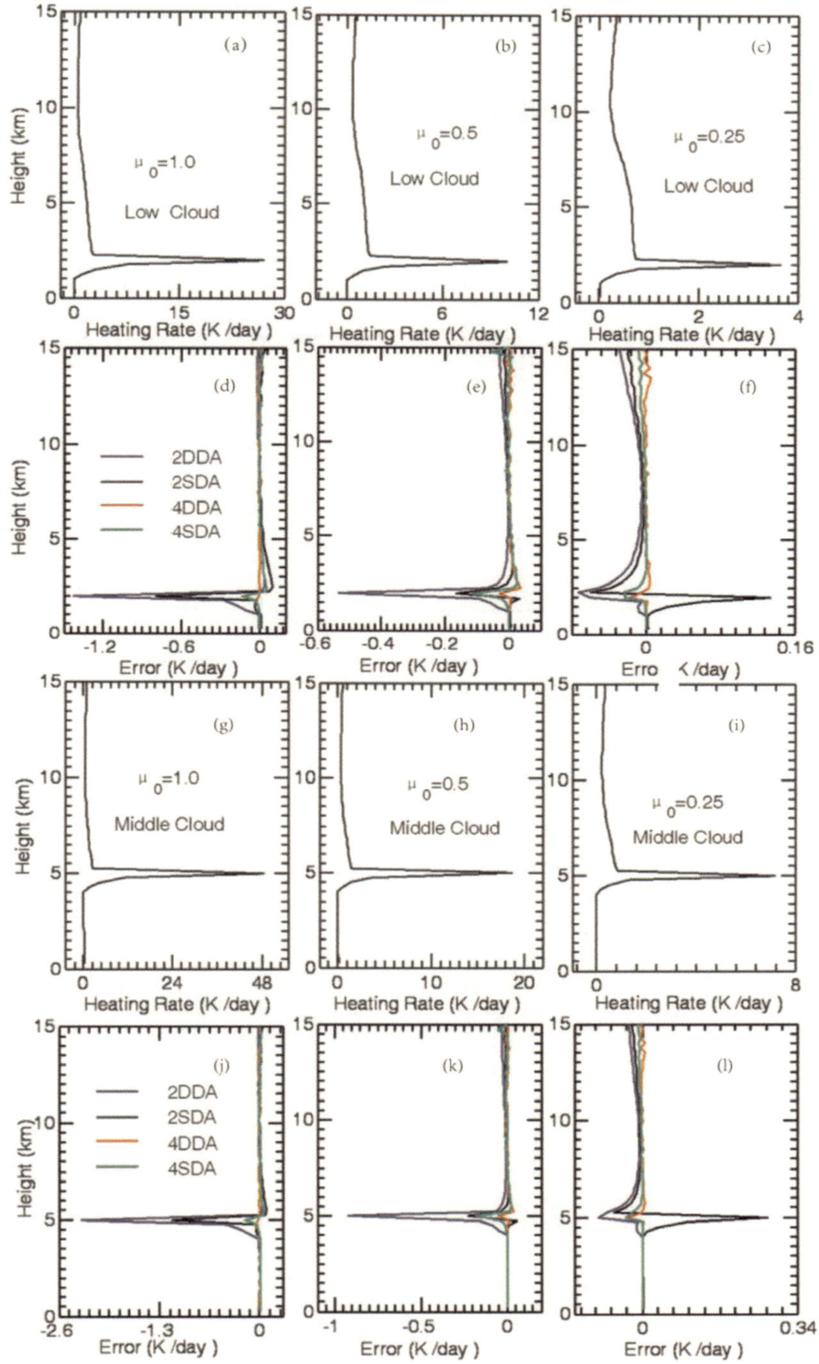

Figure 1. Heating rate profiles computed from 128S and the error profiles from 2DDA, 2SDA, 4DDA, and 4SDA. Three solar zenith angles $\mu_0 = 1$, $\mu_0 = 0.5$, and $\mu_0 = 0.25$ are considered.

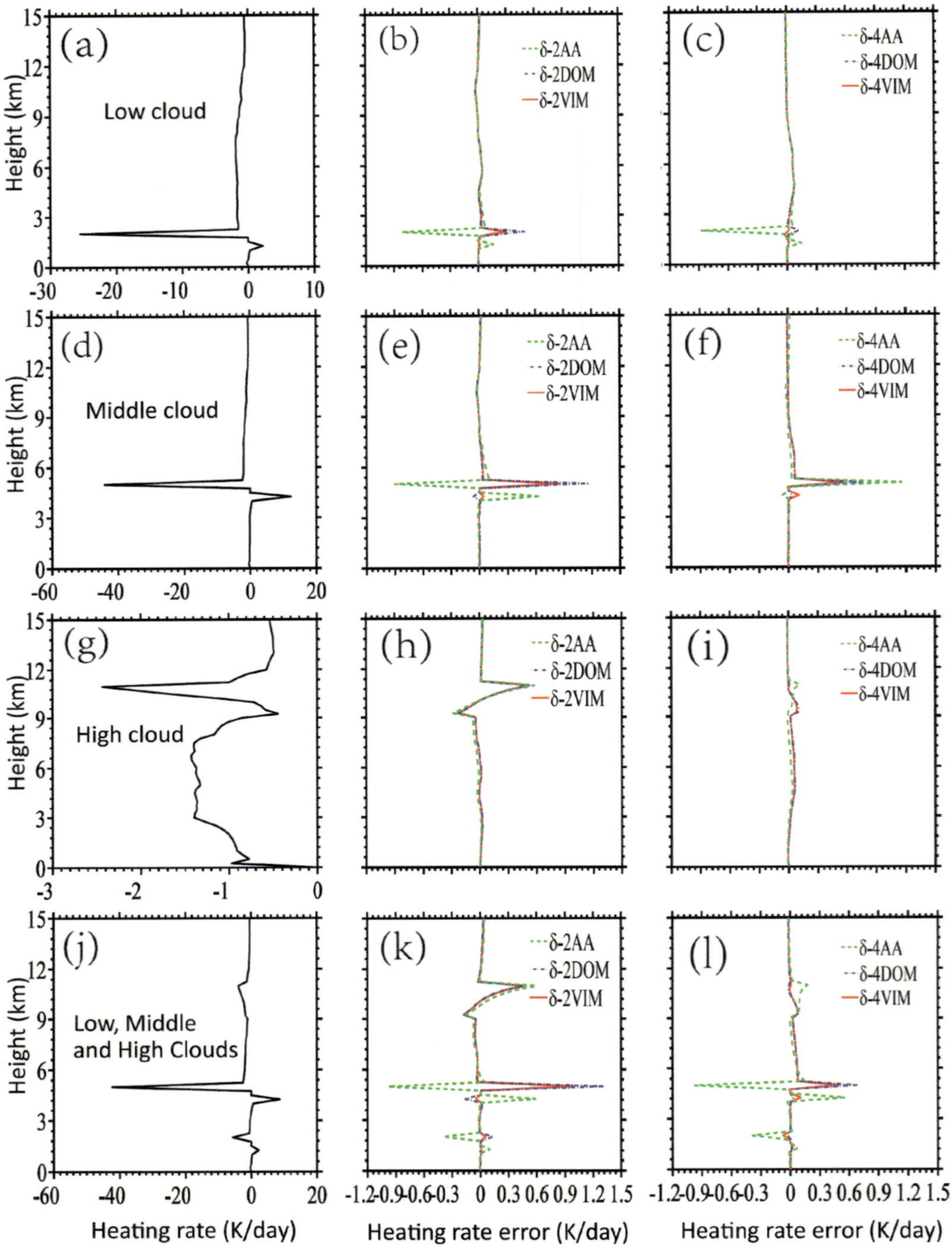

Figure 2. Heating rate profiles calculated by δ-128DOM and the error profiles produced by δ-2AA, δ-4AA, δ-2DOM, δ-4DOM, δ-2VIM, and δ-4VIM in the mid-latitude winter atmospheric profile with surface emissivity $\varepsilon_s = 1$ for (a)–(c) low cloud, (d)–(f) middle cloud, (g)–(i) high cloud, and (j)–(l) all three cloud types.

For the low-cloud case, the maximum errors of δ-2AA, δ-2DOM, and δ-2VIM are 20.8, 10.5, and 10.3 K day^{-1}, respectively, at a height corresponding to the top of the cloud (**Figure 2b**). The maximum errors of δ-4AA, δ-4DOM, and δ-4VIM are 20.9, 10.1, and 10.03 K day^{-1}, respectively, at the same height (**Figure 2c**). This shows that, for low-level water clouds, δ-2VIM (δ-4VIM) is more accurate than δ-2AA (δ-4AA) and δ-2DOM (δ-4DOM).

For the middle cloud, the maximum errors of δ-2AA, δ-2DOM, and δ-2VIM are approximately 20.9, 11.1, and 10.8 K day^{-1}, respectively, at a height corresponding to the top of the cloud. At the bottom of the cloud, the errors are 10.6 K day^{-1} for δ-2AA and just 20.05 and 10.05 K day^{-1} for δ-2DOM and δ-2VIM, respectively (**Figure 2e**). At the top of the cloud, δ-4AA, δ-4DOM, and δ-4VIM produce biases of approximately 11.1, 10.7, and 10.5 K day^{-1}, respectively (**Figure 2f**).

For the high-cloud case, the optical depth of the ice cloud is much smaller than that of the water cloud. Based on the results shown in **Figure 2**, the accuracies of δ-2DOM and δ-2VIM are similar and are better than that of δ-2AA. Furthermore, δ-4AA, δ-4DOM, and δ-4VIM have very similar accuracies. As seen in the third row in **Figure 2**, δ-2AA produces an error of approximately 10.6 K day^{-1} at a height corresponding to the top of a high cloud. The

Atmospheric condition	**128DOM**	**2AA**	**2DOM**	**2VIM**	**4AA**	**4DOM**	**4VIM**
Upward flux (TOA)							
Low clouds	209.5326	211.2735 (1.7409)	208.5206 (−1.012)	209.0425 (−0.4901)	210.9585 (1.4259)	209.1645 (−0.3681)	209.5458 (0.0132)
Middle clouds	183.5646	185.7200 (2.1554)	182.4027 (−1.1619)	182.9060 (−0.6586)	182.4002 (−1.1644)	183.1096 (−0.4550)	183.5357 (−0.0289)
High clouds	198.1357	200.3545 (2.2188)	198.6394 (0.5037)	198.8373 (0.7016)	199.2441 (1.1084)	197.2351 (−0.9006)	197.4354 (−0.7003)
Low, middle, and high clouds	176.3821	179.7768 (3.3947)	175.6421 (−0.7400)	176.2356 (−0.1465)	178.8788 (2.4967)	175.5891 (−0.7930)	175.8856 (−0.4965)
Mean square error		6.0309	0.7939	0.2969	2.7128	0.4456	0.1845
Downward flux (surface)							
Low clouds	302.7236	302.7267 (0.0031)	302.9533 (0.2297)	302.9384 (0.2148)	302.7429 (0.0193)	302.8662 (0.1426)	302.8449 (0.1213)
Middle clouds	287.4458	286.9155 (−0.5303)	287.7658 (0.3200)	287.6141 (0.1683)	287.7658 (0.3200)	287.5678 (0.1220)	287.5418 (0.096)
High clouds	247.8497	248.0313 (0.1816)	248.3309 (0.4812)	248.3303 (0.4806)	248.2968 (0.4471)	248.5608 (0.7111)	248.6452 (0.7955)
Low, middle, and high clouds	302.1410	302.0497 (−0.0913)	302.4413 (0.3003)	302.4227 (0.2817)	302.0189 (−0.1221)	302.3161 (0.1751)	302.3151 (0.1741)
Mean square error		0.0806	0.1192	0.0962	0.0794	0.1429	0.1718

The flux differences between the six approximate schemes and δ-128DOM are listed in parentheses. All δ symbols are neglected in the table.

Table 1. Comparison of δ-2AA, δ-4AA, δ-2DOM, δ-4DOM, δ-2VIM, and δ-4VIM for flux (W m^{-2}) at the top and surface by using the mid-latitude winter atmospheric profile with $\varepsilon_s = 1$.

	2AA	4AA	2DOM	4DOM	2VIM	4VIM
Algorithm only	1.0	1.4412	2.2223	14.8195	2.0365	5.3615
Radiation model	1.0	1.1095	1.4114	5.8899	1.3759	2.2739

All δ symbols are neglected in the table.

Table 2. Computing times of δ-2AA, δ-4AA, δ-2DOM, δ-4DOM, δ-2DDA, δ-4DDA, δ-2VIM, and δ-4VIM (normalized by the computing time of δ-2AA).

accuracies of δ-2DOM and δ-2VIM are similar; the maximum errors of both are approximately 10.5 K day^{-1} (**Figure 2h**). The maximum errors of δ-4AA, δ-4DOM, and δ-4VIM are all about 10.1 K day^{-1}(**Figure 2i**). However, the difference is that the maximum errors occur at top of the high cloud for δ-4AA but on the bottom for δ-4DOM and δ-4VIM.

In the case of all three clouds together, δ-2VIM is more accurate than δ-2AA and δ-2DOM for the low and middle clouds (**Figure 2k**). In general, δ-4DOM and δ-4VIM are comparable in accuracy for cloud heating rate (**Figure 2l**).

The results of upward (downward) flux at the TOA (surface) for the six schemes are presented in **Table 1** for mid-latitude winter profiles with the surface emissivity $\varepsilon_s = 1$. For the low and middle clouds, the δ-2AA overestimates the upward flux at TOA by 1.8 and 2.1 W m^{-2}, respectively. The errors of the δ-2DOM are up to 21.0 and 21.2 W m^{-2}, while that of the δ-2VIM is limited to 20.7 W m^{-2}. Also, the δ-4VIM is generally more accurate than the δ-4DOM. However, the results for the high clouds are reversed, with the δ-2VIM errors becoming larger than that of δ-2DOM. For the downward flux at the surface, the AA schemes are generally more accurate than the corresponding DOM and VIM schemes except in the middle-cloud case. We also calculate the mean square error of each scheme. For upward flux at TOA, the mean square error of VIM is generally smaller than one of DOM and AA. However, the mean square error of 2VIM and 4VIM is 0.10 W m^{-2} and 0.17 W m^{-2}, respectively, while the mean square error of 2AA is limited to 0.08 W m^{-2}.

For climate modeling, the efficiency of radiative transfer parameterization is also very important. **Table 2** lists the computing times required for δ-2AA, δ-4AA, δ-2DOM, δ-4DOM, δ-2VIM, and δ-4VIM, which were computed by HP EliteDesk 880 G1 TWR with 8 Intel(R) Core(TM) i7–4790 CPUs, 32-bit operating system, and 8GB memory. The results are normalized to that of δ-2AA. The computational efficiency of δ-2VIM is slightly better than that of δ-2DOM, which took more than double the time of δ-2AA. However, δ-4VIM is more than twice as fast as δ-4DOM for the radiation algorithm alone and the radiation model.

5. Summary and conclusions

The objective of the paper is to introduce 4DDA/4SDA [13, 20] for solar RT, AA, and VIM [32] for infrared RT and applied them to radiative transfer models.

4DDA use Gaussian quadrature method to deal with the integration in the RT equation. 4SDA is based on four-stream harmonic expansion in radiative intensity. By applying 4DDA/4SDA to a realistic atmospheric profile with gaseous transmission considered, it is found that the accuracy of 4DDA/4SDA is superior to Eddington adding method (2SDA) and two-stream discrete ordinates adding method (2DDA), especially for the cloudy conditions. It is shown that the relative errors of 4SDA are generally less than 1% in heating rate, while the relative errors of both 2SDA and 2DDA are over 6%.

VIM differs from other analytical methods for solving nonlinear differential equations in that it requires neither linearization nor small perturbations. The optimal result is constructed through variation by a Lagrange multiplier. It was found that the scattering term in the infrared RT equation could be dealt with as a nonlinear operator in VIM. By taking the AA solution as the zeroth-order solution, the scattering effect was properly included in the first-order iterative solution.

The six schemes of δ-2AA, δ-4AA, δ-2DOM, δ-4DOM, δ-2VIM, and δ-4VIM were compared systematically against the benchmark results provided by δ-128DOM for a multilayer case. For a multilayer atmosphere, VIM gave more accurate results than those of DOM and AA for the low and middle clouds in both the two- and four-stream cases. However, the errors from VIM for high clouds were similar to those from AA.

Computationally, δ-2VIM and δ-2DDA were slightly faster than δ-2DOM, which took more than double time of δ-2AA. However, δ-4VIM/δ-4DDA was more than twice as fast as δ-4DOM for both the pure radiation algorithm and the radiation model in a layered, cloudy atmosphere. In general, the main benefit of δ-2VIM was an improved accuracy with a computational time similar to that of δ-2DOM. The main benefit of δ-4VIM/δ-4DDA was improved computational efficiency with accuracy similar to that of δ-4DOM.

In view of their overall high accuracy and computational efficiency, the conclusion is that 4DDA, 4SDA, δ-2VIM, and δ-4VIM are well suited for parameterizing the solar/infrared RT in climate models.

Acknowledgements

The work is supported by the National Key R&D Program of China (2018YFC1507002) and National Natural Science Foundation of China (41675003 and 41675056).

Author details

Feng Zhang[1*], Yi-Ning Shi[1], Kun Wu[1], Jiangnan Li[2] and Wenwen Li[1]

*Address all correspondence to: fengzhang@nuist.edu.cn

1 Key Laboratory of Meteorological Disaster, Ministry of Education, Nanjing University of Information Sciences and Technology, Nanjing, China

2 Canadian Centre for Climate Modelling and Analysis, Environment and Climate Change Canada, University of Victoria, Victoria, British Columbia, Canada

References

[1] Yi-Ning S, Zhang F, Chan KL, Li W, Lin H, Duan M. An improved Eddington approximation method added delta-scaling effect for solar radiative flux calculation in a vertically inhomogeneous medium. Journal of Quantitative Spectroscopy and Radiative Transfer. 2019;**226**:40-50

[2] Feng Z, Wu K, Li J, Zhang H, Hu S. Radiative transfer in the region with solar and infrared spectra overlap. Journal of Quantitative Spectroscopy and Radiative Transfer. 2018;**219**: 366-378

[3] Yi-Ning S, Zhang F, Yan J-R, Iwabuchi H, Wang Z. The standard perturbation method for infrared radiative transfer in a vertically internally inhomogeneous scattering medium. Journal of Quantitative Spectroscopy and Radiative Transfer. 2018;**213**:149-158

[4] Feng Z, Yan J-R, Li J, Wu K, Iwabuchi H, Shi Y-N. A new radiative transfer method for solar radiation in a vertically internally inhomogeneous medium. Journal of the Atmospheric Sciences. 2018;**75**:41-55

[5] Wu K, Zhang F, Iwabuchi H, Shi Y-N, Duan M. Double-delta-function adjustment in thermal radiative transfer. Infrared Physics & Technology. 2017;**86**:139-146

[6] Meador WE, Weaver WR. Two-stream approximations to radiative transfer in planetary atmospheres: A unified description of existing methods and a new improvement. Journal of the Atmospheric Sciences. 1980;**37**:630-643

[7] Oreopoulos L et al. The continual intercomparison of radiation codes: Results from phase I. Journal of Geophysical Research. 2012;**117**:D06118

[8] Liou K-N, Fu Q, Ackerman TP. A simple formulation of the delta-four-stream approximation for radiative transfer parameterizations. Journal of the Atmospheric Sciences. 1988;**45**: 1940-1947

[9] Li J, Ramaswamy V. Four-stream spherical harmonic expansion approximation for solar radiative transfer. Journal of the Atmospheric Sciences. 1996;**53**:1174-1186

[10] Zhang H, Zhang F, Fu Q, Shen Z, Lu P. Two- and four-stream combination approximations for computation of diffuse actinic fluxes. Journal of the Atmospheric Sciences. 2010; **67**:3238-3252

[11] van Oss RF, Spurr RJ. Fast and accurate 4 and 6 stream linearized discrete ordinate radiative transfer models for ozone profile retrieval. Journal of Quantitative Spectroscopy & Radiative Transfer. 2002;**75**:177-220

[12] Lin L, Fu Q, Zhang H, Su J, Yang Q, Sun Z. Upward mass fluxes in tropical upper troposphere and lower strato-sphere derived from radiative transfer calculations. Journal of Quantitative Spectroscopy & Radiative Transfer. 2013;**117**:114-122

[13] Zhang F, Shen Z, Li J, Zhou X, Ma L. Analytical delta-four-stream doubling-adding method for radiative transfer parameterizations. Journal of the Atmospheric Sciences. 2013;**70**:794-808

[14] Chandrasekhar S. Radiative Transfer. London (Oxford University Press)

[15] Chou M-D, Suarez MJ. A solar radiation parameterization for atmospheric studies. NASA Technical Memorandum: NASA/TM-1999-104606. Vol. 15; 1999. 40pp

[16] Nakajima T, Tsukamoto M, Tsushima Y, Numaguti A, Kimura T. Modeling of the radiative process in an atmospheric general circulation model. Applied Optics. 2000;**39**: 4869-4878

[17] Li J, Barker HW. A radiation algorithm with correlated-k distribution. Part I: Local thermal equilibrium. Journal of the Atmospheric Sciences. 2005;**62**:286-309

[18] Lu C, Liu Y, Niu S, Endo S. Scale dependence of entrainment-mixing mechanisms in cumulus clouds. Journal of Geophysical Research – Atmospheres. 2014;**119**:13877-13890

[19] Lu C, Liu Y, Zhang GJ, Wu X, Endo S, Cao L, et al. Improving parameterization of entrainment rate for shallow convection with aircraft measurements and large-eddy simulation. Journal of the Atmospheric Sciences. 2016;**73**:761-773

[20] Zhang F, Li J. Doubling-adding method for delta-four-stream spherical harmonic expansion approximation in radiative transfer parameterization. Journal of the Atmospheric Sciences. 2013;**70**:3084-3101

[21] Fu Q. Parameterization of radiative processes in vertically nonhomogeneous multiple scattering atmospheres [PhD thesis]. University of Utah; 1991. 259pp

[22] Li J, Fu Q. Absorption approximation with scattering effect for infrared radiation. Journal of the Atmospheric Sciences. 2000;**57**:2905-2914

[23] Zhang F, Wu K, Li J, Zhao J-Q, Li J. Analytical infrared delta-four-stream adding method from invariance principle. Journal of the Atmospheric Sciences. 2016;**73**:4171-4188

[24] He JH. Variational iteration method for delay differential equations. Communications in Nonlinear Science and Numerical Simulation. 1997;**2**:235-236

[25] He JH. Variational iteration method—A kind of non-linear analytical technique: Some examples. International Journal of Non-Linear Mechanics. 1999;**34**:699-708

[26] He JH. Variational iteration method—Some recent results and new interpretations. Journal of Computational and Applied Mathematics. 2007;**207**:3-7

[27] Tatari M, Dehghan M. On the convergence of He's variational iteration method. Journal of Computational and Applied Mathematics. 2007;**207**:121-128

[28] Liu H, Xiao A, Su L. Convergence of variational iteration method for second-order delay differential equations. Journal of Applied Mathematics. 2013;**2013**:634670

[29] Jafari H. A comparison between the variational iteration method and the successive approximations method. Applied Mathematics Letters. 2014;**32**:1-5

[30] Al-Sawoor AJ, Al-Amr MO. Numerical solution of a reaction-diffusion system with fast reversible reaction by using Adomian's decomposition method and He's variational iteration method. AL-Rafidain Journal of Computer Sciences and Mathematics. 2012;**9**(2): 243-257

[31] Al-Sawoor AJ, Al-Amr MO. A new modification of variational iteration method for solving reaction-diffusion system with fast reversible reaction. Journal of the Egyptian Mathematical Society. 2014;**22**:396-401

[32] Zhang F, Yi-Ning S, Li J, Iwabuchi H. Variational iteration method for infrared radiative transfer in a scattering medium. Journal of the Atmospheric Sciences. 2017;**74**:419-430

[33] Stamnes K, Tsay SC, Wiscombe WJ, Jayaweera K. Numerically stable algorithm for discrete ordinate method radiative transfer in multiple scattering and emitting layered media. Applied Optics. 1988;**27**:2502-2509

[34] Wiscombe W. The delta-M method: Rapid yet accurate radiative flux calculations. Journal of the Atmospheric Sciences. 1977;**34**:1408-1422

[35] Qiang F, Liou KN. On the correlated k-distribution method for radiative transfer in nonhomogeneous atmospheres. Journal of the Atmospheric Sciences. 1992;**49**:2139-2156

[36] McClatchey RA, Fenn RW, Selby JEA, Volz FE, Garing JS. Optical properties of the atmosphere. Air Force Report: AFCRL-71-0279; 1972. 85pp

[37] Fu Q. Parameterization of radiative processes in vertically nonhomogeneous multiple scattering atmospheres [PhD dissertation]. University of Utah, 1991; 259pp [Available from: University Microfilm, 305 N. Zeeb Rd., Ann Arbor, MI 48106]

[38] Yang P, Liou K, Bi L, Liu C, Yi B, Baum B. On the radiative properties of ice clouds: Light scattering, remote sensing, and radiation parameterization. Advances in Atmospheric Sciences. 2015;**32**:32-63

[39] Lindner TH, Li J. Parameterization of the optical properties for water clouds in the infrared. Journal of Climate. 2000;**13**:1797-1805

Chapter 7

Evaluating Cooling Tower Scheme and Mechanical Drag Coefficient Formulation in High-Resolution Regional Model

Miao Yu and Shiguang Miao

Additional information is available at the end of the chapter

http://dx.doi.org/10.5772/intechopen.80522

Abstract

A cooling tower scheme considering quantitative sensible and latent heat flux released from air condition was implemented in building energy model (BEM) and coupled to the regional model (WRF). A mechanical drag coefficient formulation was implemented into the WRF/BEM to improve the representation of the wind speed in complex urban environments. Two simulations used default WRF/BEP+BEM and improved WRF/BEM to estimate the improvement effects focusing on dry day and wet day for summer 2015, respectively. The cooling tower system in commercial area not only induces the significant increase of the anthropogenic heat partition by 90% of the total heat flux releasing as latent but also further changes the surface heat flux feature. When the cooling tower is introduced, averaged surface latent heat flux in urban area is increased to about 60 $W{\cdot}m^{-2}$ with the peak of 150 $W{\cdot}m^{-2}$ in dry day and 40 $W{\cdot}m^{-2}$ with the peak of 150 $W{\cdot}m^{-2}$ in wet day. Maximum and minimum temperature error improved by 2–3 degrees. In the vertical model, the performance of boundary layer structure in rural area is much better than in urban area. The average wind speed error improved by 2–3 m/s in urban area with new calculation scheme.

Keywords: cooling tower, drag coefficient, regional model, high-resolution

1. Introduction

The world's population is coming increasingly urbanized, and most of this additional urbanization occurs in developing countries [1]. The land-use change and the anthropogenic heat release induced by urbanization have been recognized as important factors that have serious impacts on climate at regional scales [2–4]. There is plenty of evidence that the regional climate effect of urban is significant [5–9]. And urban impacts are becoming more and more important in fine weather forecasting. It is difficult to distinguish the impact of land-use change and artificial heat emissions on regional climate in the observation. But numerical model can be used to solve this problem [10]. Feng et al. [11] employed WRF coupled with single-layer urban canopy model (UCM) to investigate urban land-use change and anthropogenic heat release on regional climate in China and indicated that impact of anthropogenic heat release is larger than urban land-use change.

Anthropogenic heat is one contributor to the urban heat islands which destroyed the near-surface inversion and increased the stratification instability [12]. Anthropogenic heat release is the extra heat flux which can change the surface energy balance [13, 14]. Energy consumption from buildings is an important part of anthropogenic heat release that may modify near-surface energy balance [15].

Sailor [16] provides a historical perspective on the development of models of urban energy consumption and anthropogenic impacts on the urban energy balance. It indicated that there is a positive feedback cycle that higher temperatures result from greater amounts of energy used for air cooling in most urban area [17]. Global modeling results indicated that heat release from building is the largest contributor (89–96%) to the large-scale urban consumption of energy [12]. Future climate experiments by GCMs show anthropogenic heat flux can cause annual-mean warming of 0.4–0.9°C over large industrialized regions although global-mean anthropogenic heat flux is small [18]. The temperature increased by anthropogenic heat not only depends on the amount of heat released but also on orographical factors [19]. The amount of heat released at night is lower than at day, but the temperature increase is nearly three times greater [20]. Global model shows that the extra heat from energy consumption over the 86 major metropolitan can cause up to 1° of warming in winter seasons [21].

Recent regional modeling results show that anthropogenic heat flux from building has a significant impact on temperature simulation on urban area [22]. The heat release of air condition caused about 1–2°C warming in summer commercial area [23]. The study of Pairs also indicated that about 0.5°C results from anthropogenic heat release and points out the air condition makes important contributions to surface warming [24]. A study on three major urban agglomerations of China suggests that contribution of anthropogenic heat release to warming is larger than the land-use change [11].

But the performance of current urban canopy model is not satisfied for artificial heat emissions in urban area [11, 25]. Both UCM and BEM take the anthropogenic heat as extra surface sensible flux and can only recognize the diurnal variation [26]. But UCM and BEM cannot describe the energy exchange between anthropogenic heat flux and the urban boundary layer sensible heat flux which leads to energy balance in the boundary layer destroyed. It is noteworthy that anthropogenic heat release includes not only sensible heat flux but also latent heat flux. Anthropologic latent heat flux from urban area is ignored by UCM and BEM. Building energy model that can accurately describe energy balance mechanism of urban boundary layer and includes anthropogenic heat release urgently needs development.

The development of BEM model has solved the problem to a great extent [27]. Current BEM model is capable of describing (1) the heat diffusion through walls, roofs, and floors, (2) the natural ventilation and the radiation exchanged between the indoor surfaces, (3) the heat from occupants and equipment, and (4) the energy consumption from air condition.

Although BEM has the ability to simulate the building energy-exchange process as mentioned above, the performance is not satisfied enough especially for the high-resolution forecast in urban area [3, 25, 28]. Air condition releasing is treated as sensible heat flux to potential temperature equation when couples to BEP and regional model [29]. Simulated temperature in urban area is always obviously higher than the observation by current WRF/BEM model. Assessment report in Beijing shows that latent heat flux maximum simulated by WRF/BEM is only 40 $W{\cdot}m^{-2}$ while the observation is about 230 $W{\cdot}m^{-2}$ [25]. Errors of simulated heat flux directly lead to underestimate the humidity and further affect the performance of rainfall. In most commercial buildings, anthropogenic heat can be associated to heat release from air conditioning systems. Most air conditioning systems use evaporative cooling that releases a mixture of sensible and latent heat to the environment [30].

Many studies show that the energy consumption of air condition from building is gradually increasing as the frequency of heat wave is increasing [31]. And heat release from the building air condition system is one of the primary sources of anthropologic latent heat flux in urban area [29, 32–35]. Although heat released by air condition is considered in the current WRF/BEM, the performance is still not satisfied. Previous studies have indicated that heat released by air condition in some megacities is equal to or more than half of the surface sensible flux [23]. Simulation results show that contribution of heat released by air condition to summer warming can exceed 1° in the megacities [23, 36, 37], and contribution to nighttime temperature can reach 2° [29].

However, most air condition systems use evaporative cooling that releases a mixture of sensible and latent heat to the environment. In summer, 50–80% of their heat released by evaporative-cooled systems is in the form of latent heating [38]. In China, metropolitan electricity consumption report shows that most important energy consumption in building comes from air condition system whether in commercial or residential area [39]. Air condition usage report about Chinese metropolitan indicated that the ratio of sensible and latent heat flux by different types of air condition emissions is 20 and 80%, respectively [39].

So how to correctly describe the latent heat flux released by air condition in high-resolution model is an urgent problem to solve. It has shown that a BEM coupled with a cooling tower model can improve the model performance of temperature [30]. Cooling tower scheme will obviously improve model performance of the energy exchange ability between building and its surroundings and urban boundary layer balance.

Beijing power consumption is gradually increasing from 1978 to 2015, and the proportion of electricity consumed by residents is also gradually increasing (Beijing Municipal Bureau of Statistics website). It is noteworthy that heat released from building air condition is the important component of summer electricity consumed by residents. Namely, heat flux released by air condition system in Beijing's urban area becomes one of the primary sources of summer anthropologic heat as other big cities. In order to modify the simulation of heat flux released by air condition, a new cooling tower scheme [30] was coupled to rapid-refresh multi-scale analysis and prediction (RMAPS).

Parameterizations using the specific input parameters describe the complex arrangement of buildings and streets in an urban environment. However, simulations using this type of data reproduce fine-resolution features that are not clearly reproducible by traditional methods [10, 29].

In order to represent the effects of horizontal and vertical building surfaces of momentum, heat, and turbulent kinetic energy (TKE) equations, the building effect parameterization and the building energy model (BEP + BEM) [40] have been introduced. In the relevant equations, new terms about frictional and drag forces on the mean flow and the increase of the TKE between buildings are introduced. This scheme assumes drag coefficient is a constant. That is inaccurate because the magnitude of the drag is decided by building density in highly heterogeneous urban environments. A new formulation has been implemented in the BEP + BEM system to calculate the values of the drag coefficient based on the building plan-area fraction to improve the airflow simulation in the urban boundary layer. The performance of this drag formulation has been evaluated in an idealized urban configuration using computational fluid dynamical (CFD) simulations [27].

Our aims are as follows:

1. To improve the performance of BEM using the cooling tower scheme and drag formulation.
2. To evaluate the forecast performance of the improved RMAPS coupled with improved BEM in summer Beijing

Details of the data and the experimental design are given in Section 2. The performance of improved RMAPS is evaluated in Section 3. We summarize the findings and discuss our results in Section 4.

2. Data, model description, and experimental design

The surface temperature and humidity data were obtained from 294 meteorological stations operated by the Beijing Meteorological Bureau (**Figure 1**). Vertical temperature data was gathered from a radiometer located at 39.8°N, 116.46°E. Heat flux data used in this study were obtained from the Beijing meteorological tower (39.97°N, 116.37°E), which is 325 m high and located in North Beijing.

This study used operational rapid-refresh multi-scale analysis and prediction system (RMAPS) based on modified version of the WRF model (ARW versions 3.5.1), and its data assimilation system (WRFDA v3.5) was developed by the Institute of Urban Meteorology, China Meteorological Administration, Beijing [25]. The system starts with ECMWF global forecasts (at 3-h intervals) and terminates with hourly weather forecasts. Initial conditions is adjusted by WRFDA-ingested including S/C band weather radar, ground-based global positioning system meteorology (GPS-MET), radiosonde, Aircraft Meteorological Data Relay (AMDAR), and AWS surface observations. Three domains are designed for the current study with horizontal

http://dx.doi.org/10.5772/intechopen.80522

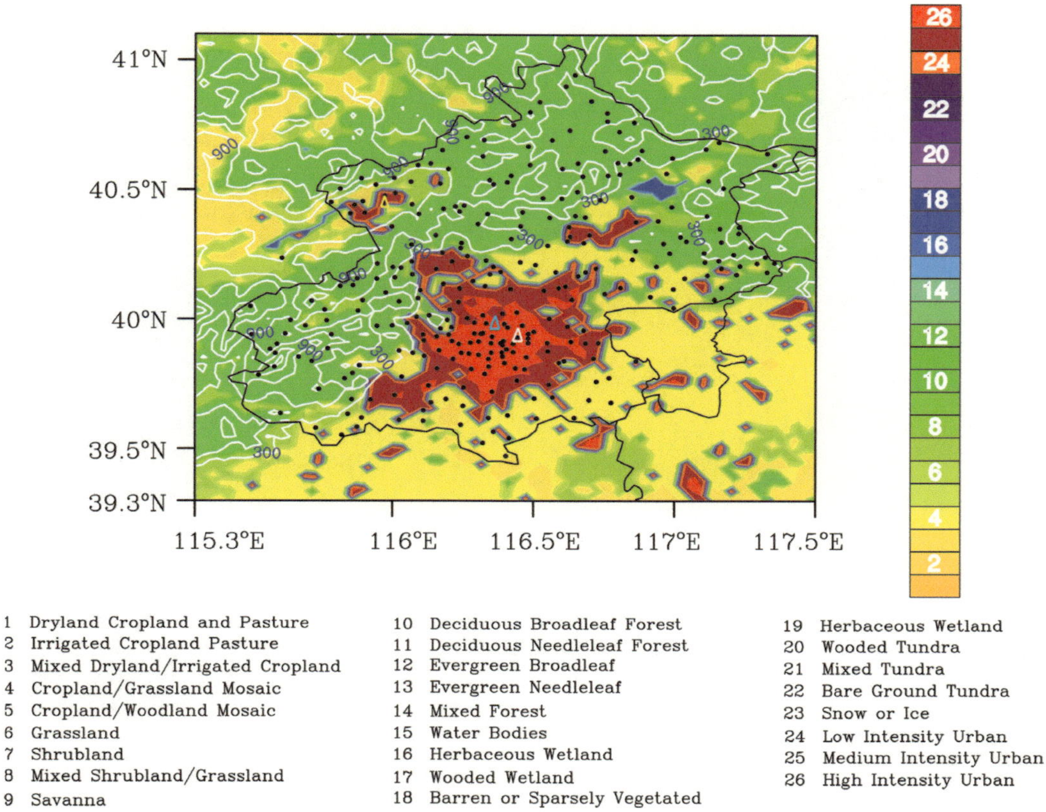

Figure 1. High-resolution land-use map in model. Black dots show the 294 weather stations. The white triangle shows the location of a 325 m meteorological tower (39.97°N, 116.37°E); the blue triangle shows the location of the Naojiao station (39.8°N, 116.46°E); and the yellow triangle shows the location of the Yanqing station (40.45°N,115.97°E).

grid spacings of 9, 3, and 1 km. The locations of the nested urban domains are shown in **Figure 2**. NDOWN provides boundary conditions for the 1-km Domain-3 model from its 3-km output, and VDRAS output is assimilated into the 1-km domain via FDDA [41]. Land-use map (**Figure 1**) is based on 30-m Landsat data for the year 2010, including three urban land types according to gridded urban-fraction values [42]. Parameterization schemes used in this study are listed in **Table 1**.

In order to improve the current forecast model, a cooling tower scheme was incorporated to the BEP + BEM and was coupled with RMAPS. Beijing is taken as the case study to investigate anthropogenic heat impact of dense urban environment. Although the cooling tower scheme has been used to the regional model in previous work, verification and evaluation for improved model are not sufficient especially in vertical stratification. This work used multiple intense observation data to evaluate the improvement effect of the new cooling tower scheme.

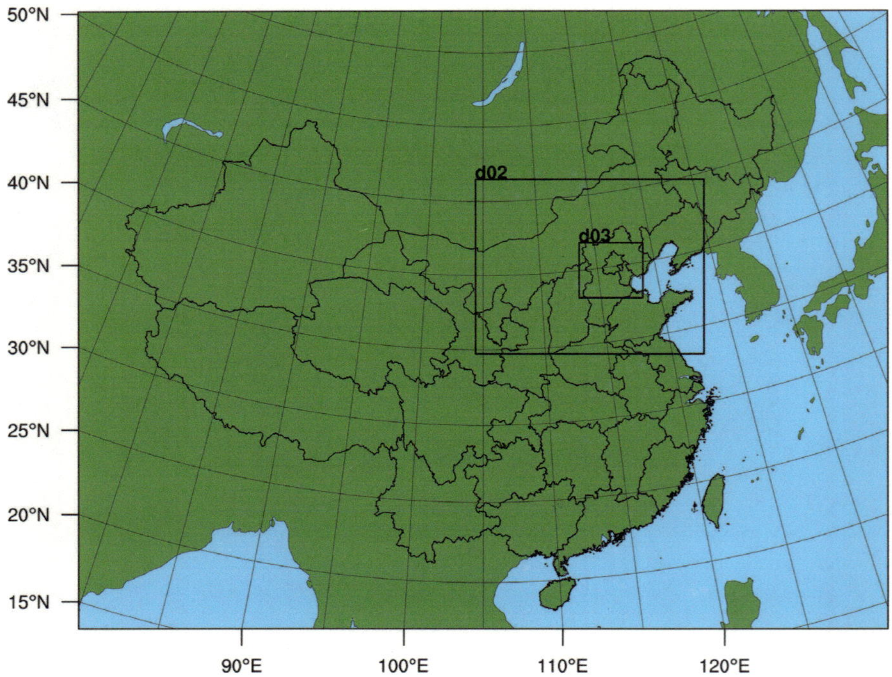

Figure 2. Domain configuration and location of the study area.

	D1	D2	D1
Models and versions	WRFDA v3.5.1 + WRF v3.5.1		WRF v3.5.1
Horizontal grid points	649 × 400	550 × 424	460 × 403
Δx (km)	9	3	1
Vertical layers	50		
Cumulus physics	Kain-Fritsch	None	None
LW radiation	RRTM		
SW radiation	Dudhia		
Microphysics	Thompson		
PBL physics	ACM2	BouLac	
Urban physics	SLUCM	BEP	BEP + BEM

Table 1. Modeling setting and parameterization options.

The computing method is as follows:

Based on the first law of thermodynamics, heat exchange equation between air condition system and the external atmosphere is defined as

$$Q = C_{min}(T_{wo,cond} - T_{wi,cond}) = C_{min}(T_{Refi} - T_{wi,cond}) \tag{1}$$

$$T_{wi,cond} = T_{wo,CT} = T_{wb,air} \tag{2}$$

$$T_{wo,cond} = T_{wi,CT} \tag{3}$$

where Q is the total heat transfer from the building calculated by BEM, $T_{wi,cond}$ is the water temperature entering the air condition system, $T_{wo,cond}$ is the water temperature leaving the air condition, $T_{wi,CT}$ is the water temperature entering the cooling, $T_{wo,CT}$ is the water temperature leaving the cooling tower (CT), and C_{min} is the minimum thermal capacitance between the water and the refrigerant [43]. It is assumed that the cooling tower is able to bring the water entering the air conditioning to its minimum value of the wet bulb temperature. This wet bulb temperature $T_{wb,air}$ is calculated from [44–48]:

$$\begin{aligned} T_{wb,air} = {} & T_{air} atan\left[0.151977(RH\% + 8.313659)^{1/2}\right] \\ & + atan\,(T_{air} + RH\%) - atan(RH\% - 1.676331) \\ & + 0.00391838(RH\%)^{3/2} atan(0.023101 RH\%) - 4.686035 \end{aligned} \tag{4}$$

where T_{air} is air temperature and RH is relative humidity.

The effectiveness, ε, for the cooling tower is defined as:

$$\varepsilon = \frac{Q}{m_a(h_{sai} - h_{ai})} \tag{5}$$

where h_{ai} is the enthalpy of inlet air and h_{sai} is saturated enthalpy of inlet air.

Finally, the outlet air temperature T_{ao} can be obtained from the following equations using an iterative scheme:

$$Q = m_a(h_{ao} - h_{ai}) \tag{6}$$

$$h_{ao} = h_{ai} + \varepsilon(h_{sai} - h_{ai}) \tag{7}$$

$$C_p T_{ao} + q_{vao}(C_{pw} + L) = h_{ao} \tag{8}$$

$$q_{vao} = 0.62198 \frac{e}{P - e} \tag{9}$$

$$e = 6.11 \times 10^{\frac{7.5 T_{ao}}{237.7 + T_{ao}}} \tag{10}$$

where q_{vao} is absolute humidity or mixing ratio.

Due to the effect of complicated urban surface, the structure of the meteorological field in the urban boundary layer is different from other surfaces. Impervious vertical surfaces of buildings induce a drag force that produces a loss of momentum that changes the flow field in near-surface atmospheric boundary layer. The drag coefficient (Cd) is an important component for calculating the magnitude of the momentum flux induced by buildings in urban canopy models. According to previous studies over urban environments and wind-tunnel measurements of, Cd is assumed as a constant (0.4) in default WRF/BEM [40]. However, Cd could vary with building packing densities. An analytical relation proposed by Salamanca and Martilli [27] has been implemented into the BEP + BEM system to estimate the drag coefficient as a function of the building plan-area fraction as follows:

$$\text{Cdeq}\,(\lambda\text{p}) = \begin{cases} 3.32\text{x}\ \lambda\text{p}^{0.47} & \textit{for}\ \lambda\text{p} \le 0.29 \\ 1.85 & \textit{for}\ \lambda\text{p} > 0.29 \end{cases} \tag{11}$$

This formulation represents an improvement compared to using a constant drag coefficient, and it is necessary to assess this for a real complex urban underlay.

We evaluated the whole summer (from June 1 to September 30) simulation to evaluate the performance of the RMAPS coupled to the cooling tower model and drag scheme (AC + VD). The forecast results by default RMAPS were used as control run (CTL).

3. Results

3.1. Effect on diurnal pattern

The significant difference between CTL and AC + VD is in latent heat released from the building because AC + VD improved the heat flux released from the building to the environment. Maximum sensible heat flux from air condition in CTL is about 180 $W{\cdot}m^{-2}$, while it is reduced to 20 $W{\cdot}m^{-2}$ in AC + VD (**Figure 3a** and **b**). Meanwhile latent heat flux from air condition is increased by AC + VD during daytime. Thermal exchange between building and

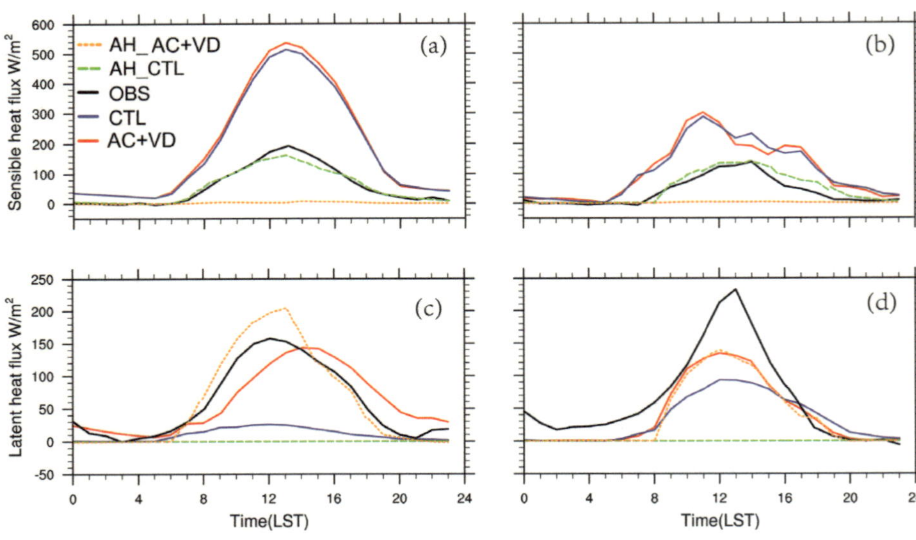

Figure 3. Averaged diurnal pattern of heat flux ($W{\cdot}m^{-2}$). (a) Sensible heat flux in dry day; (b) sensible heat flux in wet day; (c) latent heat flux in dry day; and (d) latent heat flux in wet day.

its external atmosphere simulation is an inadequate capability in CTL. AC + VD not only increase the latent heat flux released from building but also improve simulation ability of heat exchange.

We evaluate sensible and latent heat flux in the dry day first. Based on the heat flux observation by Beijing tower in 140 m, sensible heat flux is less than 20 $W{\cdot}m^{-2}$ in the nighttime, while CTL overestimate sensible heat flux about 50 $W{\cdot}m^{-2}$ in the nighttime (**Figure 3a**). It has largely solved this problem by AC + VD during the dry days. But in the wet days, sensible heat flux is still overestimated in nighttime simulated by both CTL and AC + VD. In the daytime, observed sensible heat flux reaches the maximum (200 $W{\cdot}m^{-2}$) at 1400 LST; both CTL and AC + VD overestimate the sensible heat flux in urban area (**Figure 3a** and **b**). Sensible heat flux simulated by CTL delays the time to reach the maximum of about 1 hour while an hour earlier by AC + VD. Both CTL and AC + VD overestimate sensible heat flux in urban area, but AC + VD improve the simulation results from 1500 to 2000 LST especially in the dry days. Compared to dry day result, improvement effect for sensible heat flux by ECs is not obvious in wet day. Different from the dry day, the sensible heat flux is rapidly decreasing during the 1100–1800 LST in model results because rainfall often occurs in that duration. And simulated sensible heat flux is more sensitive to precipitation than observed.

Observed latent heat flux in dry days has the same features as the sensible heat flux in daytime (**Figure 3c**). The maximum of observed latent heat flux is about 200 $W{\cdot}m^{-2}$, while the latent heat flux is seriously underestimated by CTL in urban area which further leads to error in humidity and temperature. But AC + VD result indicates that model performance for latent heat flux is much improved in both dry day and wet day (**Figure 3c** and **d**). The simulated diurnal pattern of latent heat flux by AC + VD is very close to the observation in the dry days although the value is still less than observation and there is phase deviation. There are still errors of latent heat flux simulated by AC + VD in the wet days.

Bowen ratio is a very important index for the energy balance. Because the latent heat flux is seriously underestimated by CTL, there are big simulation errors for Bowen ratio especially in nighttime, both in dry days and wet days (**Figure 4**). That problem has been largely solved by AC + VD even though there are still underestimated Bowen ratios especially from 600 to 1600 LST.

The heat flux change by AC + VD will further influence temperature and humidity in urban area. So next part we will force on evaluating the model performance of temperature and humidity at 2 m. For the dry day, temperature is obviously overestimated by CTL during the

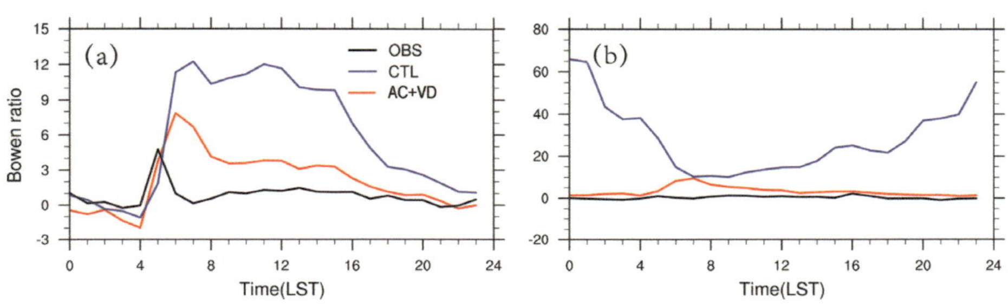

Figure 4. Averaged diurnal pattern of Bowen ratio. (a) Dry day and (b) wet day.

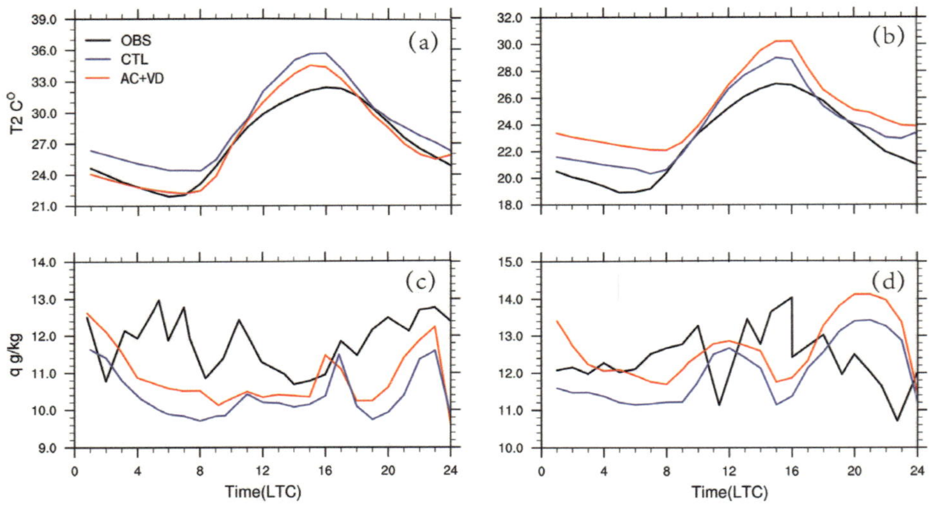

Figure 5. Averaged diurnal pattern of temperature (°C) and absolute humidity ($g{\cdot}kg^{-1}$). (a) Temperature in dry day; (b) temperature in wet day; (c) absolute humidity in dry day; and (d) absolute in wet day.

whole day. But performance is largely improved by AC + VD especially in nighttime. Temperature at 2 m is decreased around 3° by AC + VD in nighttime which is very close to the observation (**Figure 5a**). There are still obvious errors during 1200–1600 LST. For the dry days, improvement for nighttime is still significant (2°) but not as good as wet days when compared to the observation (**Figure 5b**). Although AC + VD improves the temperature about 1.5°, the maximum simulation deviation still occurred from 1200 to 1600 LST. That is related to overestimating the sensible heat flux. Another error of phase is still in the simulation in dry day.

For the dry day, absolute humidity is underestimated by CTL during the whole day, while it is improved by AC + VD especially in nighttime (**Figure 5c**). For the wet day, absolute humidity is increasing from 1200 to 1600 because rainfall will more likely occur in this period. While simulated absolute humidity in both CTL and AC + VD lags behind the observation, increasing period is from 1800 to 2300 LST (**Figure 5d**). Simulated value is improved by AC + VD although the phase difference still remained in both dry day and wet day.

3.2. Effect on spatial distribution

294 meteorological stations are used to evaluate model performance of temperature and humidity spatial distribution. Spatial distribution of averaged temperature error shows that CTL overestimate daily mean temperature in most of urban station (**Figure 6a**), and the errors of most stations reach 1–2°. Errors of temperature are obviously reduced by EC in urban area and errors of about half stations <0.5° (**Figure 6b**). Both CTL and AC + VD underestimate daily mean absolute humidity at most urban stations. And there is no significant difference or improved effect by AC + VD in mean absolute humidity in urban area.

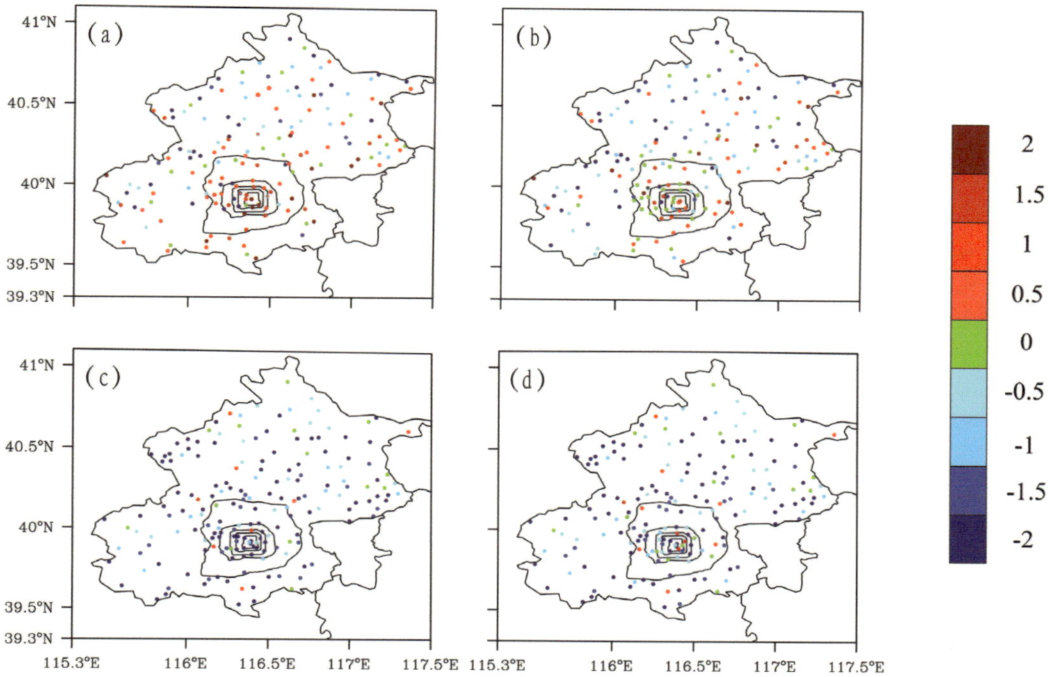

Figure 6. Spatial distribution of averaged temperature (°C) and absolute humidity (g·kg^{-1}) errors (difference between modeling and observation) at 2 m, the black circles show the second to sixth ring roads. (a) Temperature error in CTL; (b) temperature error in AC + VD; (c) absolute humidity error in CTL; and (d) absolute humidity error in AC + VD.

Maximum temperature (at 3 pm) simulated by CTL is more than 35°C in urban and suburban area in the dry day and 33°C in most of the plains (**Figure 7a**). While temperature is reduced to about 2° in AC + VD in both urban and suburban areas, and the area in which temperature is more than 36°C is obviously decreased (**Figure 7b**). There are the same characteristics for wet days (**Figure 7c** and **d**).

Spatial distribution of maximum sensible heat flux shows that sensible heat flux simulated by CTL in urban area is more than 350 W·m^{-2} in the dry day (**Figure 8a**). The sensible heat flux simulated by AC + VD in most urban areas is about 320 W·m^{-2} which is a little smaller than CTL (**Figure 8b**) in the dry day. And the maximum region (more than 350 W·m^{-2}) of sensible heat flux is also reduced by AC + VD. In the wet day, sensible heat flux maximum simulated by AC + VD is smaller about 50–100 W·m^{-2} than CTL (**Figure 8c** and **d**).

Latent heat flux maximum simulated by CTL is less than 50 W·m^{-2} in downtown area which is obviously underestimated compared to the observation in both dry day and wet day (**Figure 9a** and **c**). Simulation of latent heat flux in urban area by AC + VD improves the value to 100 W·m^{-2} in dry day and 150 W·m^{-2} in wet day (**Figure 9b** and **d**).

Maximum sensible heat flux released by air condition in CTL is more than 100 W·m^{-2} in urban area, and there is a little difference between dry day and wet day. AC + VD reduce the maximum sensible heat flux released by the building's air condition in both dry days and wet

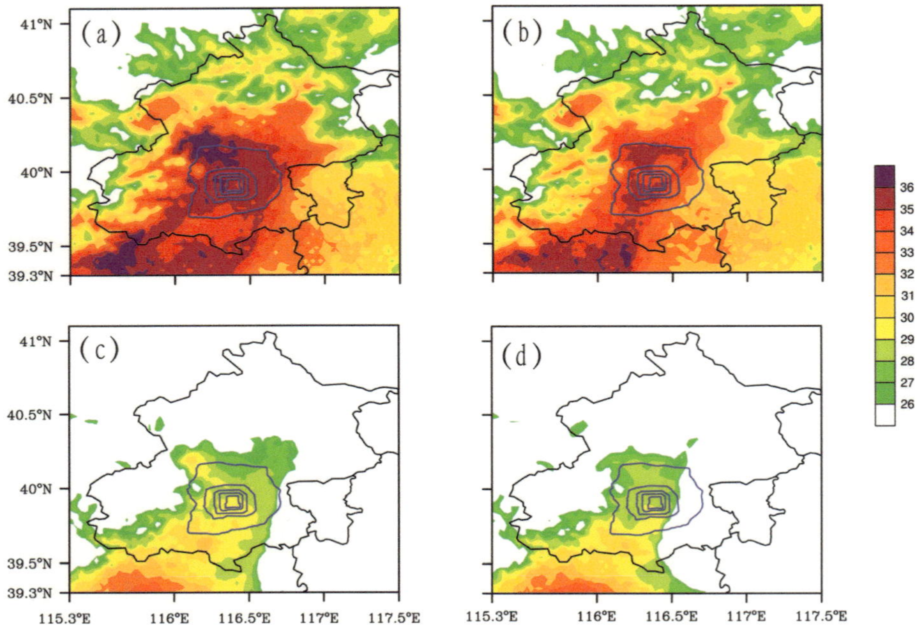

Figure 7. Spatial distribution of averaged temperature (°C) at 2 m in 3 pm. (a) Simulated by CTL in dry day; (b) simulated by AC + VD in dry day; (c) simulated by CTL in wet day; and (d) simulated by AC + VD in wet day.

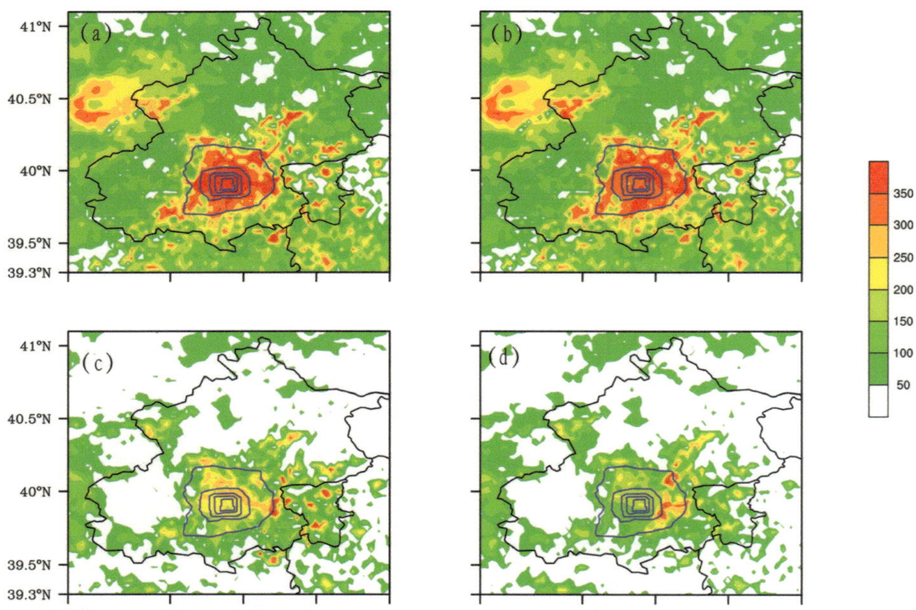

Figure 8. Spatial distribution of averaged sensible heat flux (W·m^{-2}) in 3 pm. (a) Simulated by CTL in dry day; (b) simulated by AC + VD in dry day; (c) simulated by CTL in wet day; and (d) simulated by AC + VD in wet day.

http://dx.doi.org/10.5772/intechopen.80522

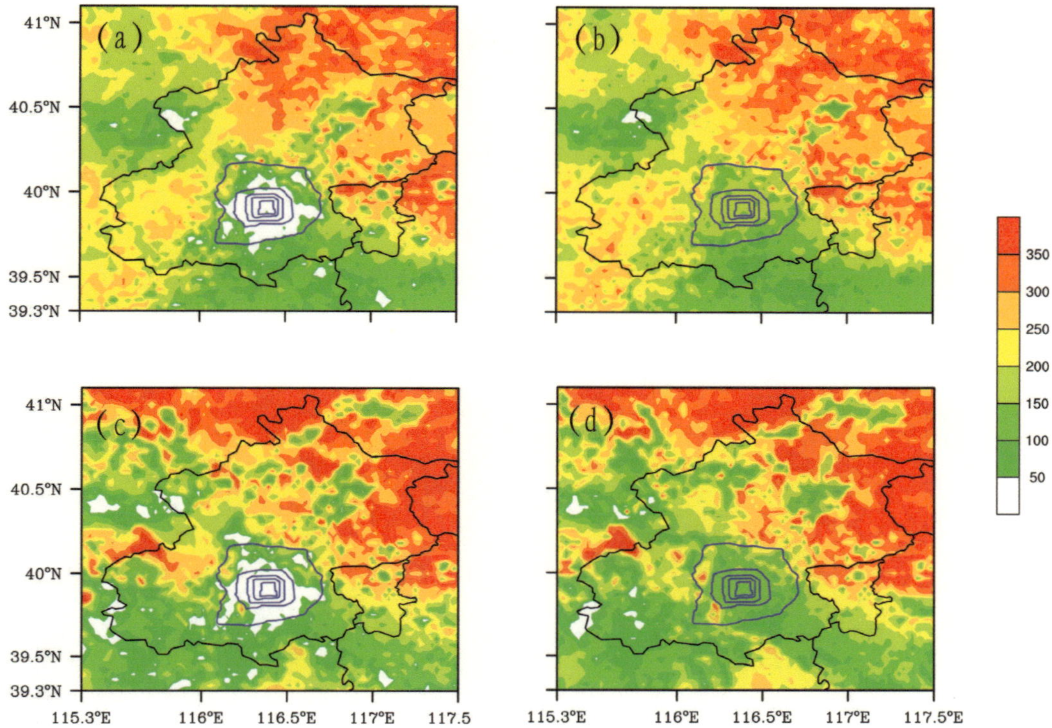

Figure 9. As in **Figure 8** but for latent heat flux.

days (**Figure 10a** and **c**). And sensible heat flux in dry day is obviously larger than wet days simulated by AC + VD (**Figure 10b** and **d**). That means sensible heat flux released by air condition in AC + VD is affected by outdoor temperature. Model performance of indoor and outdoor exchange is significantly improved by AC + VD.

There is no latent heat flux released by air conditioning used in the potential temperature equation when coupled to the WRF/BEP + BEM in CTL. However, in AC + VD the maximum latent heat flux released is more than 120 $W{\cdot}m^{-2}$ in dry days and 80–100 $W{\cdot}m^{-2}$ in wet days (**Figure 11a** and **b**) over urban core areas.

3.3. Effect on wind field

The observed daytime change of urban area averaged wind speed appeared as single peak. The wind speed reaches maximum and minimum at 700 LST and 000 LST in summer in Beijing (**Figure 12a**). The diurnal feature of wind speed is well captured by CTL and AC + VD. However, CTL overestimated wind speed in daytime especially during 500–1000 LST. This problem has been largely solved by AC + VD (**Figure 12a**). RMSE is also obviously reduced by AC + VD in the whole day (**Figure 12b**).

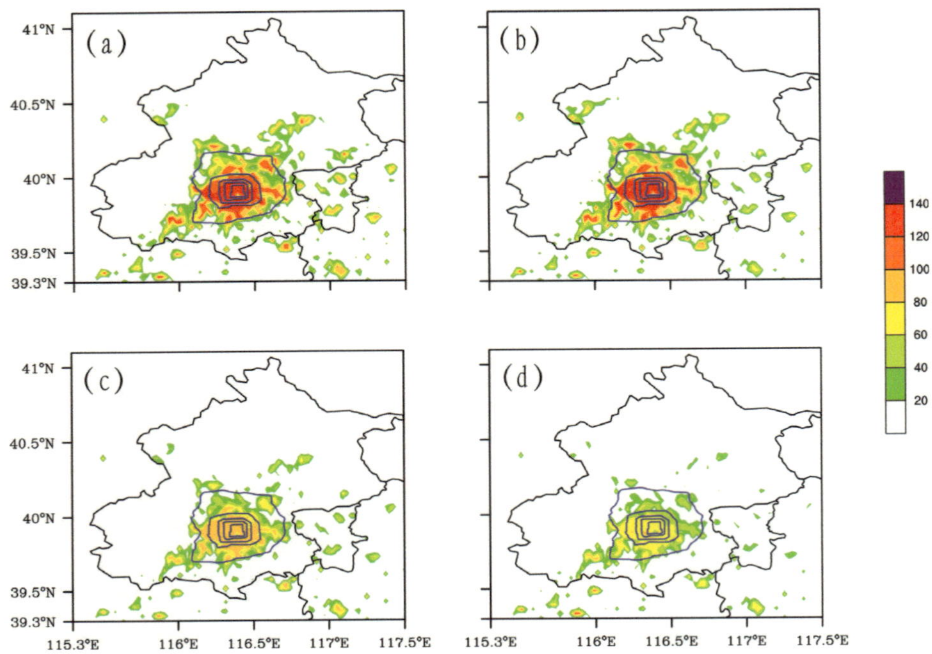

Figure 10. Spatial distribution of sensible heat flux (W·m^{-2}) released by building air condition in 3 pm. (a) Simulated by CTL in dry day; (b) simulated by AC + VD in dry day; (c) simulated by CTL in wet day; and (d) simulated by AC + VD in wet day.

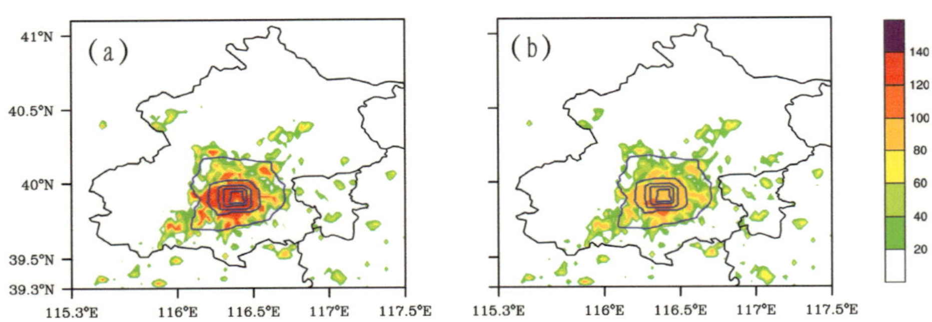

Figure 11. Spatial distribution of latent heat flux (W·m^{-2}) released by building air condition at 3 pm. (a) Simulated by AC + VD in dry day and (b) simulated by AC + VD in wet day.

Spatial distributions of wind speed by CTL show that the average wind speed in urban area is 3.5 W·m^{-2} (**Figure 13a**), while the wind speed is reduced in AC + VD about 1.5 W·m^{-2} (**Figure 13b**). The averaged wind speed simulated by AC + VD is about 1.8 W·m^{-2}. Wind speed in CTL is overestimated in all Beijing areas (**Figure 14a**). Spatial distributions of wind speed errors also indicated that wind speed error is obviously revised by AC + VD (**Figure 14b**).

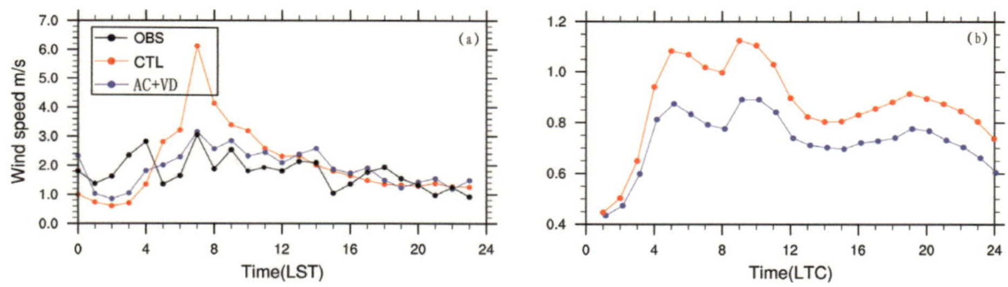

Figure 12. Averaged diurnal pattern of wind speed and its RMSE (W·m^{-2}). (a) Averaged of study period and (b) RMSE.

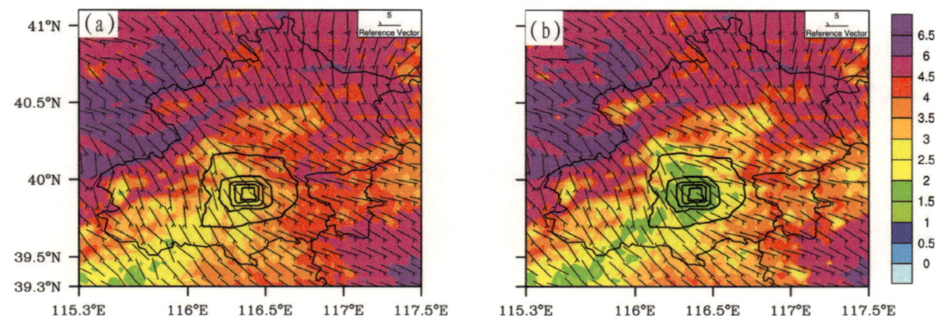

Figure 13. Spatial distributions of wind speed and wind fields (W·m^{-2}). (a) CTL and (b) AC + VD.

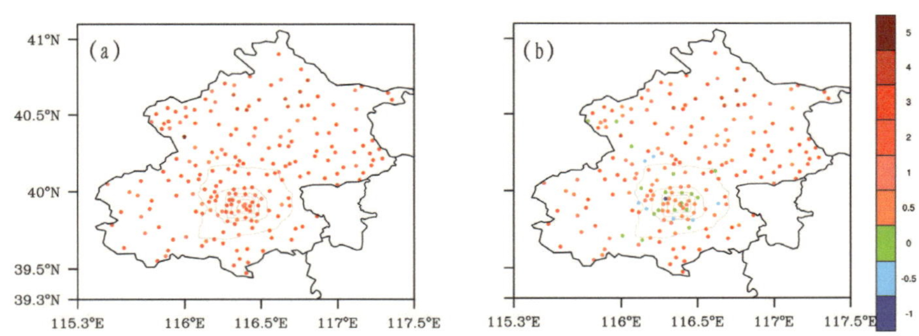

Figure 14. Spatial distributions of wind speed errors (m·s^{-1}) (difference between modeling and observation) at 10 m. (a) CTL and (b) AC + VD.

4. Conclusions and discussions

A cooling tower scheme considering quantitative sensible and latent heat flux released from air condition was coupled to RMAPS. A mechanical drag coefficient formulation was implemented into the RMAPS to improve the representation of the wind speed in complex urban environments. The computing method is based on the heat transfer between temperature and humidity and between condenser and outdoor inlet air. Two simulations use default RMAPS and improved RMAPS to estimate the improvement effect focusing on dry day and wet day, respectively. The cooling tower system in commercial area not only induces the significant increase of the anthropogenic heat partition by 90% of the total heat flux releasing as latent but also further changes the surface heat flux feature. When cooling tower is introduced, averaged surface latent heat flux in urban area is increased to about 60 $W{\cdot}m^{-2}$ with the peak of 150 $W{\cdot}m^{-2}$ in dry day and 40 $W{\cdot}m^{-2}$ with the peak of 150 $W{\cdot}m^{-2}$ in wet day. Further new cooling tower scheme improves the model performance of temperature and humidity. Maximum and minimum temperature error improves 2–3° especially in dry day. The drag coefficient formulation induced the simulated wind speed to about 2.5 $m{\cdot}s^{-1}$ that improve the wind speed error of about 2–3 $m{\cdot}s^{-1}$ in urban area.

Acknowledgements

This work was supported by National Natural Science Foundation of China (grant 41705090).

Author details

Miao Yu[1,2]* and Shiguang Miao[1]

*Address all correspondence to: yumiao0926@126.com

1 Institute of Urban Meteorology, China Meteorological Administration, Beijing, China

2 State Key Laboratory of Severe Weather, Chinese Academy of Meteorological Sciences, Beijing, China

References

[1] Population Reference Bureau. World Population Data Sheet. Population Reference Bureau; 2010

[2] Li Q, Li Q, Zhang H, Liu X, et al. Urban heat island effect on annual mean temperature during the last 50 years in China. Theoretical and Applied Climatology. 2004;**79**:165-174

[3] Li Y, Zhu L, Zhao X, Li S, Yan Y. Urbanization impact on temperature change in China with emphasis on land cover change and human activity. Journal of Climate. 2013;**26**:8765-8780

[4] Roth M. Review of urban climate research in (sub) tropical regions. International Journal of Climatology. 2007;**27**:1859-1873

[5] Hu Y, Dong W, He Y. Impact of land surface forcings on mean and extreme temperature in eastern China. Journal of Geophysical Research. 2010;**115**(D19)

[6] Kalnay E, Cai M. Impact of urbanization and land-use change on climate. Nature. 2003; **423**(6939):528-531

[7] Kalnay E, Cai M, Li H, Tobin J. Estimation of the impact of land-surface forcings on temperature trends in eastern United States. Journal of Geophysical Research. 2006;**111**(D6)

[8] Miao SG, Chen F, Lemone MA, et al. An observation al and modeling study of characteristics of urban heat island and boundary layer structures in Beijing. Journal of Applied Meteorology and Climatology. 2009;**48**(3):484-501

[9] Zhou LM, Dickinson RE, Tian YH, et al. Evidence for a significant urbanization effect on climate in China. Proceedings of the National Academy of Sciences of the United States of America. 2004;**101**:9540-9544

[10] Chen F, Kusaka H, Bornstein R, et al. The integrated WRF/urban modelling system: Development, evaluation, and applications to urban environmental problems. International Journal of Climatology. 2011;**31**(2):273-288

[11] Feng J, Wang Y, Ma Z, Liu Y. Simulating the regional impacts of urbanization and anthropogenic heat release on climate across China. Journal of Climate. 2012;**25**:7187-7203

[12] Allen L, Lindberg F, Grimmond CSB. Global to city scale urban anthropogenic heat flux: Model and variability. International Journal of Climatology. 2010;**31**:1990-2005

[13] Hinkel KM, Nelson FE. Anthropogenic heat island at Barrow, Alaska, during winter: 2001–2005. Journal of Geophysical Research. 2007;**112**:D06118

[14] Ichinose T, Shimodozono K, Hanaki K. Impact of anthropogenic heat on urban climate in Tokyo. Atmospheric Environment. 1999;**33**:3897-3909

[15] Zheng S, Liu S. Urbanization effect on climate in Beijing. Climate and Environmental Research. 2008;**13**(2):123-133

[16] Sailor D. A review of methods for estimating anthropogenic heat and moisture emissions in the urban environment. International Journal of Climatology. 2011;**31**:189-199

[17] Crutzen PJ. New directions: The growing urban heat and pollution "island" effect-Impact on chemistry and climate. Atmospheric Environment. 2004;**38**:3539-3540

[18] Flanner MG. Integrating anthropogenic heat flux with global climate models. Geophysical Research Letters. 2009;**36**:L02801

[19] Block A, Keuler K, Schaller E. Impacts of anthropogenic heat on regional climate patterns. Geophysical Research Letters. 2004;**31**:L12211

[20] Narumi D, Kondo A, Shimoda Y. Effects of anthropogenic heat release upon the urban climate in a Japanese megacity. Environmental Research. 2009;**109**:421-431

[21] Zhang G, Cai M, Hu A. Energy consumption and the unexplained winter warming over northern Asia and North America. Nature Climate Change. 2013;**3**:466-470

[22] He X, Jiang W, Chen Y, Liu G. Numerical simulation of the impacts of anthropogenic heat on the structure of the urban boundary layer. Chinese Journal of Geophysics. 2007;**50**(1): 75-83

[23] Ohashi Y, Genchi Y, Kondo H, Kikegawa Y, et al. Influence of air-conditioning waste heat on air temperature in Tokyo during summer: Numerical experiments using an urban canopy model coupled with a building energy model. Journal of Applied Meteorology and Climatology. 2007;**46**:66-81

[24] Munck C, Pigeon G, Masson V, Meunier F, et al. How much can air conditioning increase air temperatures for a City Like Paris, France? International Journal of Climatology. 2013; **33**(1):210-227

[25] Fan S. Assessment Report of Regional High Resolution Model (BJ-RUCv3.0). IUM Technical Note IUM/2015-1. Beijing, China: IUM; 2015

[26] Kusaka H, Kondo H, Kikegawa Y, Kimura F. A simple single-layer urban canopy model for atmospheric models: Comparison with multi-layer and slab models. Boundary-Layer Meteorology. 2001;**101**(3):329-358

[27] Salamanca F, Martilli A. A new building energy model coupled with an urban canopy parameterization for urban climate simulations–part II, Validation with one dimension off-line simulations. Theoretical and Applied Climatology. 2010;**99**:345-356

[28] Barlage M, Miao S, Chen F. Impact of physics parameterizations on high-resolution weather prediction over two Chinese megacities. Journal of Geophysical Research – Atmospheres. 2016;**121**:4487-4498

[29] Salamanca F, Martilli A, Tewari M, Chen F. A Study of the urban boundary layer using different urban parameterizations and high-resolution urban canopy parameters with WRF. Journal of Applied Meteorology and Climatology. 2011;**50**:1107-1128

[30] Gutiérrez E, Gonzalez J, Martilli A, Bornstein R, Arend M. Simulations of a heat wave event in New York City using a multilayer urban parameterization. Journal of Applied Meteorology and Climatology. 2015;**54**(2):283-301

[31] Adnot J. Energy Efficiency and Certification of Central Air Conditioners, Study for the D. G. Transportation-Energy (DGTREN) of the Commission of the EU. Ed. ARMINES. Final Report. Vol 1; April 2003. pp. 1-54

[32] Akbari H, Konopacki S. Energy savings of heat-island reduction strategies in Toronto, Canada. Energy. 2004;**29**:191-210

[33] Akbari H, Pomerantz M, Taha H. Cool surfaces and shade trees to reduce energy use and improve air quality in urban areas. Solar Energy. 2001;**70**(3):295-310

[34] Hassid S, Santamouris M, Papanikolaou N, Linardi A, Klitsikas N, Georgakis C, Assimakopoulos DN. The effect of the Athens heat island on air conditioning load. Energy and Buildings. 2000;**32**:131-141

[35] Kolokotroni M, Giannitsaris I, Watkins R. The effect of the London urban heat island on building summer cooling demand and night ventilation strategies. Solar Energy. 2006;**80**: 383-392

[36] Hsieh C-M, Aramaki T, Hanaki K. The feedback of heat rejection to air conditioning load during the nighttime in subtropical climate. Energy and Buildings. 2007;**39**:1175-1182

[37] Wen Y, Lian Z. Influence of air conditioners utilization on urban thermal environment. Applied Thermal Engineering. 2009;**29**:670-675

[38] Sailor DJ, Brooks A, Hart M, Heiple S. A bottom-up approach for estimating latent and sensible heat emissions from anthropogenic sources. In: Proceedings of the 7th Symposium on the Urban Environment. San Diego, CA: American Meteorological Society; 2007

[39] Zheng Y, Miao S, Zhang Q, et al. Improvements of building energy model and anthropogenic heat release from cooling system. Plateau Meteorology. 2015;**36**(2):562-574

[40] Martilli A, Clappier A, Rotach MW. An urban surface exchange parameterization for mesoscale models. Boundary-Layer Meteorology. 2002;**104**:261-304

[41] Zhang Y, Miao S, Dai Y, Bornstein R. Numerical simulation of urban land surface effects on summer convective rainfall under different UHI intensity in Beijing. Journal of Geophysical Research – Atmospheres. 2017;**122**(15):7851-7868

[42] Zhang Y, Miao S, Dai Y, Liu YH. Numerical simulation of characteristics of summer clear day boundary layer in Beijing and the impact of urban underlying surface on sea breeze [in Chinese]. Chinese Journal of Geophysics. 2013;**56**(8):2558-2573

[43] González JE, Bula-Silvera AJ. Conventional mechanical systems for efficient heating, ventilating, and air conditioning systems. In: Handbook of Integrated and Sustainable Buildings Equipment and Systems. Volume I: Energy Systems. New York: ASME Press; 2017

[44] Kusaka H, Kimura F. Coupling a single-layer urban canopy model with a simple atmospheric model: Impact on urban heat island simulation for an idealized case. Journal of the Meteorological Society of Japan. 2004;**82**:67-80

[45] Liang X, Miao S, Li J, et al. Surf: Understanding and predicting urban convection and haze. Bulletin of the American Meteorological Society. 2018;**99**(7):1391-1413

[46] Liao J, Tang L, Shao G, Qiu Q, Wang C, Zheng S, Su X. A neighbor decay cellular automata approach for simulating urban expansion based on particle swarm intelligence. International Journal of Geographical Information Science. 2014;**28**(4):720-738

[47] Olivo Y, Hamidi A, Ramamurthy P. Spatiotemporal variability in building energy use in New York City. Energy. 2017;**141**:1393-1401

[48] Yang B, Zhang Y, Qian Y. Simulation of urban climate with high-resolution WRF model: A case study in Nanjing, China, Asia-Pac. Journal of the Atmospheric Sciences. 2012;**48**(3): 227-241

Chapter 8

Numerical Air Quality Forecast over Eastern China: Development, Uncertainty and Future

Guangqiang Zhou, Zhongqi Yu, Yixuan Gu and Luyu Chang

Additional information is available at the end of the chapter

http://dx.doi.org/10.5772/intechopen.79304

Abstract

Air pollution is severely focused due to its distinct effect on climate change and adverse effect on human health, ecological system, etc. Eastern China is one of the most polluted areas in the world and many actions were taken to reduce air pollution. Numerical forecast of air quality was proved to be one of the effective ways to help to deal with air pollution. This chapter will present the development, uncertainty and thinking about the future of the numerical air quality forecast emphasized in eastern China region. Brief history of numerical air quality modeling including that of Shanghai Meteorological Service (SMS) was reviewed. The operational regional atmospheric environmental modeling system for eastern China (RAEMS) and its performance on forecasting the major air pollutants over eastern China region was introduced. Uncertainty was analyzed meanwhile challenges and actions to be done in the future were suggested to provide better service of numerical air quality forecast.

Keywords: numerical prediction, numerical forecast, air pollution, air quality, eastern China

1. Introduction

China has been suffering severe air pollution in recent years, characterized as high levels of fine particles ($PM_{2.5}$) and ozone [1–4]. As part of atmospheric composition, air pollutants play important role in climate change. For example, ozone is one of major greenhouse gases, which causes atmospheric warming [5]. Atmospheric aerosol is one of the most important and uncertain factors in both climate change and weather activities. It influences climate by its direct radiative forcing and induced cloud adjustments and weather by the interactions of aerosol-radiation, aerosol-cloud, etc. [5]. Air pollution also leads to adverse effects on health [6, 7], including increasing of respiratory and cardiovascular diseases, excess mortality, and decreasing of life expectancy [8–11]. High particulate matter (PM) concentration under relatively high relative humidity (RH) conditions often induces haze events and causes high risk on public activities such as surface transportation, aircraft take-off and landing. Therefore, the characteristics, formation mechanisms, and influence factors of air pollution and related issues were seriously focused in recent years (e.g. in [4, 12–15]).

In policy decision aspect, the Chinese government therefore has issued series of actions to reduce air pollution in the last few years. The new Chinese national ambient air quality standards (CNAAQS2012) [16] was jointly released by Ministry of Environmental Protection (MEP) of the People's Republic of China and General Administration of Quality Supervision, Inspection and Quarantine of the People's Republic of China in 2012. At the first time, standards for $PM_{2.5}$ and daily maximum 8-hour averaged (DM8H) ozone (O_3-8h) were established in China. The State Council then issued a stringent action plan to combat air pollution on September, 2013 [17]. China sponsored tens of projects and funded several billions since 2016 in a special fund named Study on Formation Mechanism of Atmospheric pollution and Control Technology. In the support of the Premier Fund, "2 + 26" cities were chosen and one scientific team was organized for each city in 2017 to deal with the air pollution in Beijing-Tianjin-Hebei and its surrounding region. Accordingly, China Meteorological Administration (CMA) established operational centers in three populated regions (Beijing-Tianjin-Hebei, Yangtze River Delta, and Pearl River Delta) to provide air quality forecasting and warning. Provincial governments took many kinds of actions to try to improve ambient air quality.

Eastern China, which covers the Yangtze River Delta, is one of the most polluted regions [1, 3]. The air quality in this region is also influenced by Beijing-Tianjin-Hebei region by the northwesterly. Study on air pollution as well as its secondarily produced haze in this region was thus widely carried out and numerical modeling played an important role. For example, Tie et al. studied ozone [18] and Zhou et al. studied particulate matter and haze [19] over Shanghai by using the Weather Research and Forecasting model coupled with Chemistry (WRF-Chem) [20]; the severe PM pollution and haze episodes over eastern China in January 2013 were modeled by using the nested air quality prediction model system (NAQPMS) [21] and revised Community Multi-scale Air Quality (CMAQ) model [22, 23], etc. In the previous studies, increase of secondary aerosols was certified to take important role in heavy PM pollution events (e.g. in [19, 23, 24]) and some new sources through heterogeneous processes were found to promote rapid increase of PM in extreme pollution episodes [14, 25]. These works proved that the usage of air quality models is one valid solution to air pollution studies.

In this chapter, the numerical forecast of air quality over eastern China is presented. This work is one of the important applications of numerical meteorological prediction and supports air quality and relevant service including temporary emission control and study of air pollution on health, etc. In the next sections, the brief history of development of numerical modeling for air pollution will be reviewed. Then the operational forecast will be emphasized, including the construction of modeling system and forecast performance. Analysis and discussion on the uncertainty and shortage in current work will be presented to help improving the forecast in the future. Brief conclusion will be given in the end.

2. Brief history of numerical air quality modeling

Air quality models are tools that describe the physical and chemical processes which influence air pollutants, including chemical reactions, transport, diffusion, scavenging, etc. in the atmosphere. They are built based on the understanding of atmospheric physics and chemistry and computation technology. The models are used in many air quality and related issues, such as analyzing the characteristics of tempo-spatial patterns and changes of air pollutants, discovering the mechanisms of formation of air pollution, and estimating the influence of the change of factors (e.g. anthropogenic emission, volcanic explosion) on air quality, etc. Usually, air quality models are more or less driven by meteorological variables and therefore are connected with meteorological models or model outputs.

Since the 1970s, three generations of air quality models have been developed sponsored by United States Environmental Protection Agency (US EPA) and other organizations. In the first-generation models, atmospheric physical processes are highly parameterized and chemical processes are ignored or just simply treated. These models introduce the dispersion profiles in different levels of discretized stability and are specialized in calculating the long-term average concentration of inert air pollutant. The second-generation models include more complicated meteorological models and nonlinear chemical reactions and the simulation domain is three-dimensionally (3-D) gridded. The chemical and physical processes are individually calculated in each grid and influence between neighbor grids is considered. This generation is used generally to treat one type of air pollution, such as photochemical smog and acid rain. In the end of the 1990s, US EPA presented the concept of "one atmosphere" and developed the third-generation air quality modeling system—Medels-3/CMAQ [26]. It is an integrated system and consists of serial modules to process emissions, meteorology inputs, chemical reaction and transport, production making, etc. The third-generation models involve relatively detailed atmospheric chemistry and physics as well as the influence and inter-conversion among air pollutants of different types or phases. In fact, the divide of different generations is not distinct and some models are still in continuous development. For example, the CALPIFF (one Lagrangian model of the first-generation) introduced much research results in the 1990s and was often implemented in the 2000s. The second version regional acid model (RADM2) increased chemical species and reactions [27] and was introduced in the very newly developed third-generation model of WRF-Chem [20].

In recent years, 3-D chemical transport models (CTMs) has been widely used in studying and forecasting air quality combined with numerical meteorological models benefited from the rapid development of models and computing technology. For example, global ozone was simulated by using the model for ozone and related chemical tracers (MOZART) and the model performance was evaluated [28, 29]. Gu et al. studied summertime ozone and nitrate aerosol in upper troposphere and lower stratosphere (UTLS) over the Tibetan Plateau and the south Asian monsoon region using the Goddard Earth Observing System chemical transport model (GEOS-Chem) [30, 31]. The CMAQ model had a great number of applications around the world, e.g. in [32, 33]. Tie et al. studied the characterizations of chemical oxidants in Mexico City using WRF-Chem [34]. Zhou et al. developed an operational mesoscale sand and dust storm forecasting system for East Asia by coupling a dust model within the CMA unified atmospheric chemistry environment (CUACE) [35]. Zhou et al. developed the CUACE for aerosols (CUACE/Aero) to study chemical and optical properties of aerosol in China [36]. Over eastern China, there were also numerous applications of CTMs. Gao et al. studied regional haze events in the North China Plain (NCP) using WRF-Chem [37]. Zhou et al. built an operational system to forecast air quality over eastern China region and resulted good performance in forecasting the major air pollutants of $PM_{2.5}$ and ozone over this region [38]. Wu et al. analyzed the source contribution of primary and secondary sulfate, nitrate, and ammonium (S-N-A) during a representative winter period in Shanghai using online source-tagged NAQPMS [39]. Li et al. investigated ozone source by using the ozone source apportionment technology (OSAT) with tagged tracers coupled within Comprehensive Air Quality Model with Extensions (CAMx) [40].

Air quality modeling in current generation can be switched "offline" or "online" depending on the treatment of meteorology and chemistry. The offline chemical processes are treated independently from the meteorological modeling, while those in online approach are dependent. The modeling systems implemented in recent years are mostly offline, such as AIRPACT [32]. The chemical transport in this approach is driven by outputs from a separate meteorological model, typically available once per hour. This approach is computationally attractive since only one meteorological dataset can be used to produce many chemical simulations for different scientific questions. On the other hand, the "online" treatment (e.g. WRF-Chem) was newly developed to solve the loss of information in offline approach about atmospheric processes that have a time scale of less than the output time interval of meteorological models, including wind speed, wind direction, rainfall, etc. The lost information may be very important in high resolution air quality modeling. The online approach also benefits to investigate the interactions between meteorology and chemistry [21], which are out of the purpose of offline treatments. Previous studies (e.g. in [19, 21, 37, 38, 41]) on air pollution and related issues over eastern China region had proved the applicability and advantage of the online model of WRF-Chem.

Shanghai Meteorological Service (SMS), as well as the East China Meteorological Center of CMA, shares the responsibility to provide air quality forecast and air pollution warning for Shanghai and guidance for East China region. Therefore, SMS initialized numerical modeling of air quality in 2006. This work got scientific and technological supports from the World Meteorological Organization (WMO) through Shanghai WMO global atmosphere watch (GAW) urban research and meteorological environment (GURME) Pilot Project. Based on the thinking of the applicability and advantage of WRF-Chem and the extendibility on calculation of the inter-feedback between meteorological variables and air pollutants, WRF-Chem was chosen as the core model in developing our numerical air quality forecast system. An experimental forecasting system was established in 2008, in which nested domains of 16 × 16-km and 4 × 4-km was implemented. The outer domain covered eastern China region and the inner one covered the main YRD region. The evaluation showed that the results from two domains had comparable performance and further study in [34] showed that the 6 × 6-km resolution performed best under the conditions of the model and emission data at that time. Therefore, a real time forecast system covering the YRD region with a horizontal resolution of 6-km was built in 2009 to support the air pollution (including three variables of PM_{10}, SO_2 and NO_2) forecast for Shanghai. This application showed that the forecasts from this version had acceptable performance under relatively stable conditions but poorer performance for transport cases, because there are much more air pollutants transported from areas outside the

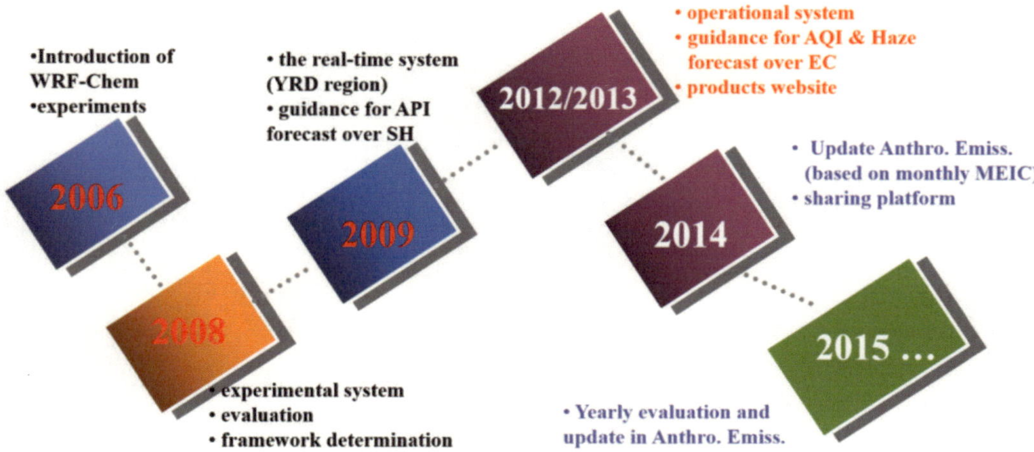

Figure 1. The brief history of development of numerical air quality forecast in SMS.

model region such as the NCP. With updates in high performance computational resource, one forecast system covering eastern China region was established in 2012, which was named as Regional Atmospheric Environmental Modeling System for eastern China (RAEMS). This system was certificated as an official operational forecast system by CMA in March, 2013. More details about the operational system will be introduced in the next section and the brief history of its development was shown in **Figure 1**.

3. The operational forecast and performance of RAEMS

3.1. Framework of the operational system

The core model in RAEMS is WRF-Chem, which was developed through the collaboration of several institutes (e.g. NOAA, NCAR, etc.). Chemistry and meteorology is fully coupled in this model, in other words, the same advection, convection, and diffusion scheme, model grids, physical schemes, and time step is used and there is no interpolation in time for meteorological fields. The modeling performance of WRF-Chem has been extensively validated [20, 42]. Several real-time prediction systems were built based on the WRF-Chem model to provide air quality forecasts around the world (e.g. China, the United States, and Brazil), as listed in [43]. In RAEMS, several improvements were made based on WRF-Chem version 3.2 by Tie et al. [44], including the introduction of aerosol effects on photolysis, adjustments of nocturnal ozone losing, and introduction of ISORROPIA II secondary inorganic aerosol scheme [45]. This modified version has been validated, showing good performance in ozone and $PM_{2.5}$ prediction for Shanghai [18, 19].

As shown in **Figure 2**, the domain encompasses the eastern China Region. Centered at (32.5°N, 118°E), it consists of 360 un-staggered grids in west-east and 400 in south-north with a 6-km grid resolution. There are 28 layers vertically, with the top pressure of 50 hPa. The time step for integration is 30-s for meteorology and 60-s for chemistry, and these for radiation, biogenic

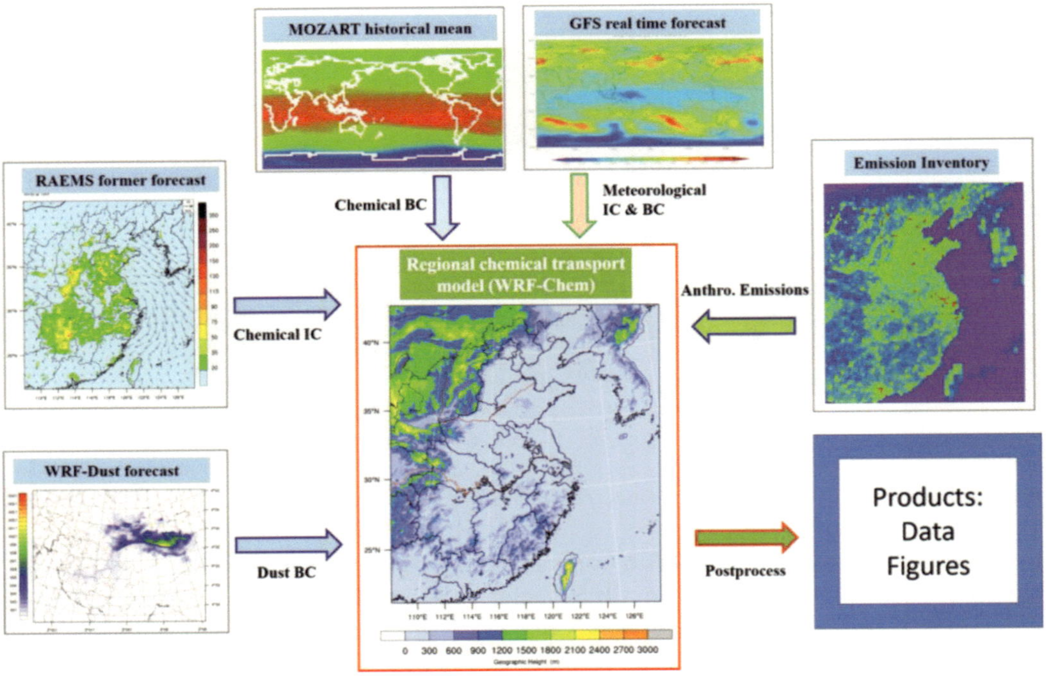

Figure 2. Components of RAEMS. Domain coverage was shown in the central component.

emission, and photolysis are 10, 30, and 15 min, respectively. Physical options are listed in **Table 1**. Specially, the Noah-modified 20-category IGBP-MODIS instead of 24-category USGS land-use was used. The RADM2 [27] was used for gas-phase chemistry. ISORROPIA II

Parameterization scheme	Option
Micro-physics (mp_physics)	WSM 6-class
Cumulus parameterization (cu_phy)	Not used
Long-wave radiation (ra_lw)	RRTM
Short-wave radiation (ra_sw)	Dudhia
Surface layer (sf_sfclay)	Monin_Obukhov
Land surface (sf_surface)	Unified Noah
Boundary layer (bl_pbl)	YSU
Gas-phase chemistry	RADM2
Inorganic aerosol chemistry	ISORROPIA II
Organic aerosol chemistry	SORGAM

Table 1. Physical and chemical configuration in RAEMS.

secondary inorganic [45] and the Secondary ORGanic Aerosol Model (SORGAM) [46] schemes were used to treat aerosol chemistry. Madronich scheme [47, 48] was applied for photolysis.

The global forecast from the National Centers for Environmental Prediction Global Forecast System (NCEP GFS) was used for meteorological initial and boundary conditions. NCEP GFS data was used widely for weather forecast, analysis, and as the initial and lateral boundary conditions of regional modeling. 0.5-degree GFS forecast was used, and 1-degree data was also applied if higher resolution forecasts were not available. Previous forecast was used for chemical initial conditions. The gaseous chemical lateral boundary conditions were based on estimations from a global chemical transport model (MOZART-4) [28, 29]. Boundary conditions were extracted from the MOZART-4 by matching the RAEMS boundary with the MOZART cells. While maintaining diurnal variations in species concentrations, monthly averaged MOZART-4 values of the year 2009 were applied.

3.2. Biogenic and anthropogenic emissions

Biogenic emissions were calculated online using model of emissions of gases and aerosols from nature (MEGAN2, in [49, 50]). Global land cover maps including isoprene emission factor, plant functional type, and leaf area index were applied.

The multi-resolution emission inventory for China (MEIC [51, 52]) for the year 2010 was applied as the anthropogenic inventory. MEIC inventory was developed by Tsinghua University, including emissions of 10 major atmospheric pollutants and greenhouse gases (SO_2, NO_X, CO, NMVOC, NH_3, CO_2, $PM_{2.5}$, PM_{10}, BC, and OC) over mainland China. MEIC supplied gridded monthly emissions from five sectors (industry, power, residential, transport, and agriculture) with a 0.25-degree resolution. Asian emission inventory for the NASA INTEX-B Mission [53] was applied for regions outside mainland China and before August, 2014. It has a resolution of 0.5-degree for the year 2008.

While being used in RAEMS system, the emissions were spatially regridded to the model grids. Emissions were also hourly allocated with the diurnal profile (in [38]) provided by Shanghai Academy of Environmental Science. NO emission took a proportion of 90% of the amount of NO_x in mole number and NO_2 took the rest 10% (as in [41]). Information of spatial distribution and total amount can be found in [38].

3.3. Operational execution and products

The RAEMS was authorized as an official operational forecasting system by CMA on Mar. 23, 2013 and has been producing forecast since then. The operational system runs once per day, initialized at 12Z UTC (20Z LST). It is started at about 2 am at local time every day and completes entire simulation and post-processing within 5 h. The predictable time length is more than 78 h and the forecast system provides forecast products for 3 local days.

Operational products are displayed on a website [54]. The link to this site is also accessible from the official NOAA WRF-Chem website [43]. The products include hourly spatial distributions of major pollutants and air quality related meteorological conditions. Temporal variations of both meteorological elements and pollutant species at more than 500 stations as well as real-time evaluation results are also provided online.

3.4. Regular update on anthropogenic emission

The anthropogenic emission used in RAEMS was yearly updated since 2016 to fit the change of emission as well as the adaptability of the modeling system. The emission was updated monthly based on that used in the same month of previous year. These adjustments were majorly depended on the results of monthly evaluation of previous year and information of emission regulation and control implementing in that month as well as the feedback from the forecasters in operational agencies who use the products every day. In the treatment, the ratio of bias median to observational average for each city was taken as the key indicator for adjusting. At the same time, performance of NO_2 and SO_2 and primary PM emission was most focused because of the importance of S-N-A in secondary aerosol [55, 56] and that of primary aerosol. For example, the evaluation showed that NO_2 was obviously underestimated in the northern and southern parts of East China region with bias ratios of over −25% in December, 2015 (**Figure 3**). SO_2 forecasts showed more serious underestimation for most cities in these two areas. But the RAEMS overestimated NO_2 and SO_2 for many cities in the middle region, especially for the cities along the Yangtze River. Therefore, the emitting intensities of NO_2 and SO_2 in December, 2016 were increased or decreased in different amounts separately for different areas. Accordingly, other emitting species were adjusted in the similar way. The amounts were estimated experientially based on ratios and control information.

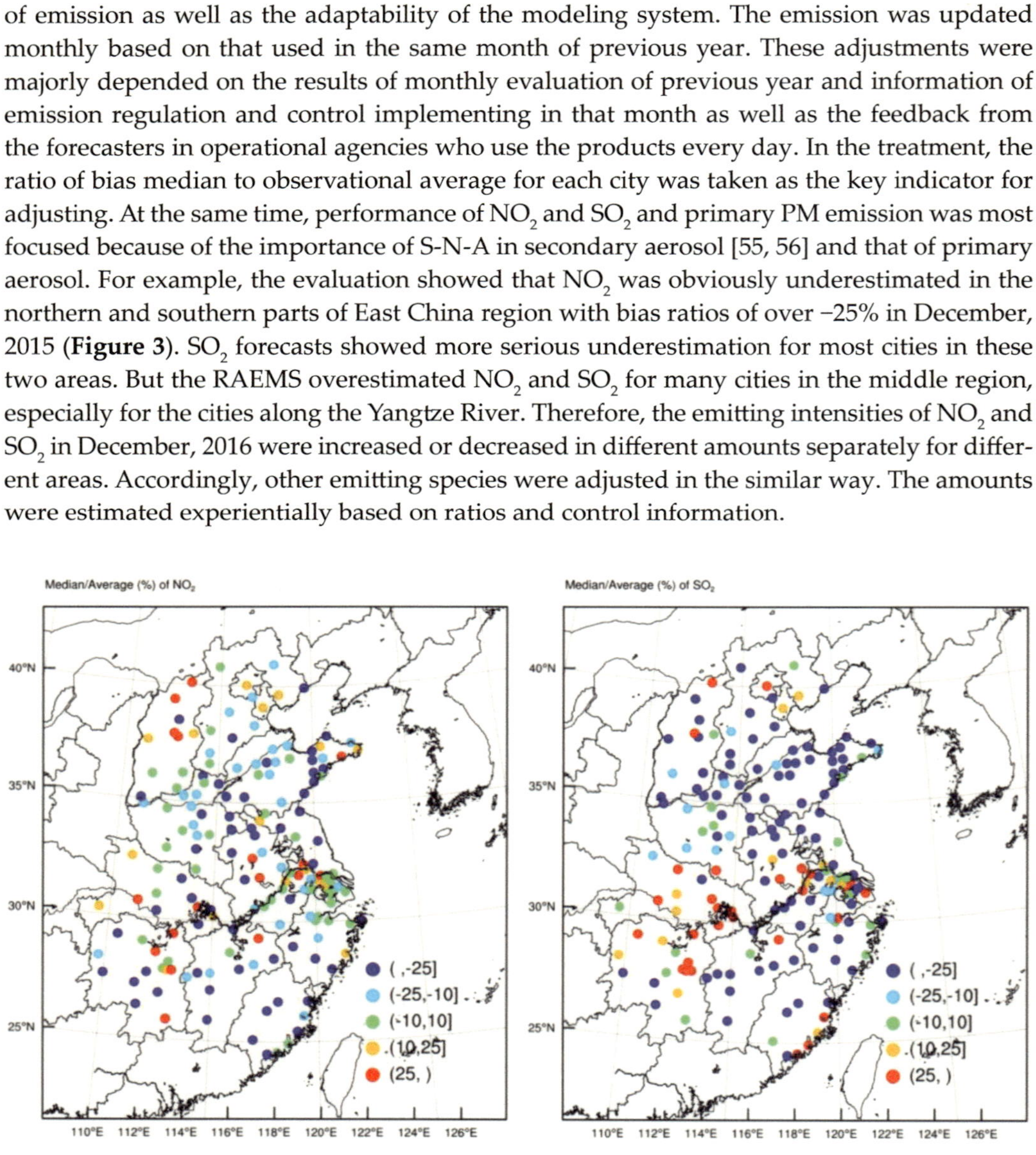

Figure 3. The distribution of the ratio of forecast bias median to the observational average in December, 2015 for NO_2 (left) and SO_2 (right).

3.5. Evaluations

A comprehensive evaluation on the performance of RAEMS was carried out in [38]. In that work, the performance in the beginning of two natural years of 2014 and 2015 was exhibited. They analyzed the series of statistical indicators for variables of $PM_{2.5}$, ozone, PM_{10}, NO_2, SO_2 and CO. The indicators included mean bias (MB), mean error (ME), root mean square error (RMSE), correlation coefficient (R), normalized mean bias (NMB) and error (NME), factor of 2 of measurement values (FAC2, the ratio of forecast records within between half and twice of measurement values), Fractional bias (FB) and error (FE), etc. Category performance with different exceedance limits was also evaluated for the two most important pollutants of $PM_{2.5}$ and O_3-8h. In spatial, the performance of $PM_{2.5}$ and DM8H ozone for main cities and $PM_{2.5}$ for provincial capital cities was shown. In temporal, the consistency of different forecast time length of $PM_{2.5}$ and ozone and diurnal variation and the distribution of peak time of ozone was analyzed.

In general, their results showed that the RAEMS has good performance in forecasting the temporal trend and spatial distribution of major air pollutants over eastern China region and the performance is consistent with the increasing forecast time length up to 3 days. All summarized statistical indicators of daily $PM_{2.5}$ and DM8H ozone in different forecast time lengths were comparable with each other and no distinct disagreements were shown. About half of cities have correlation coefficients greater than 0.6 for $PM_{2.5}$ and 0.7 for DM8H ozone. The forecasted $PM_{2.5}$ concentrations were generally in good agreements with observed concentrations, with most cities having NMB within ±25%. Forecasted ozone diurnal variation was very similar to the observations and made small peak time error. The modeling system also exhibited acceptable performance for the other air pollutants. More detailed information can be found in [38].

Here more evaluation results were given for the city of Shanghai, one of the largest cities around the world, to show a glimpse on the continuity of forecast performance and how the forecast system performed after 2015. **Figure 4** shows the scattering results of observed and 48-h forecasted $PM_{2.5}$ and O_3-8h for 4 years from 2014 to 2017. It shows that RAEMS had generally good performance in forecasting the two most important air pollutants. For $PM_{2.5}$, the four-year average observed concentration was 46.9 μg/m^3 and the forecasted concentration was only 0.1 μg/m^3 overestimated. The correlation coefficient between observation and prediction of $PM_{2.5}$ was 0.74. It also revealed relatively low RMSE and NMB, 22.3 μg/m^3 and 8.1%, respectively and high FAC2 of 0.89. This result suggested that 89% forecasted $PM_{2.5}$ concentrations were within between half and twice of those of observed. These indicators showed excellent performance in forecasting and modeling $PM_{2.5}$. The NMB of 8.1% was much lower than the acceptable threshold value of ±20% recommended in the United Kingdom [57]. For example, Chen et al. reported a FAC2 of around 60% and NMB of 17 and 32% for polluted and clean periods [32]. Grell et al. reported a R^2 of 0.38 for simulating $PM_{2.5}$ over New Hampshire using WRF-Chem [20]. Foley et al. Reported a NMB of 19% [33]. Prank et al. found underestimation of 10–60% over Europe using four chemical transport models of CMAQ, EMEP, LOTOS-EUROS and SILAM [58]. Wu et al. reported FAC2 of 70–80% [39]. For O_3-8h, the forecasts showed better performance in indicators of correlation coefficient, NMB, and FAC2, but worse in MB and RMSE comparing with corresponding indicators for $PM_{2.5}$. The performance for Shanghai has high scores among the cities over the eastern China [38].

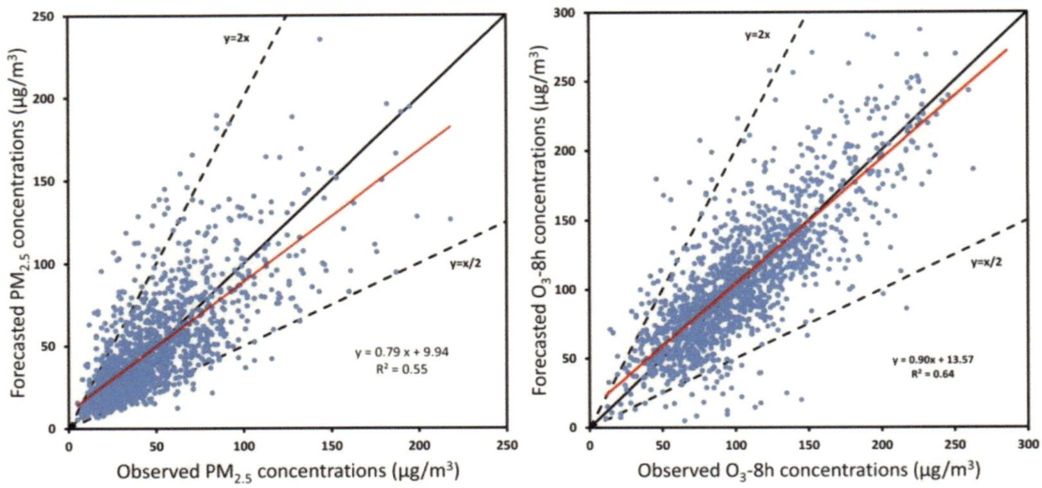

Figure 4. Scattering plot of 48-h forecasted and observed daily mean $PM_{2.5}$ and O_3-8h for shanghai during 2014–2017.

The performances for different years were generally consistent for both $PM_{2.5}$ and O_3-8h (**Table 2**). For example, the values of FAC2 were around 0.89 for $PM_{2.5}$ and 0.93–0.97 for O_3-8h, respectively. RMSEs were within 20.8–23.9 μg/m³ for $PM_{2.5}$ and 28.2–32.9 μg/m³ for O_3-8h, respectively. Correlation coefficients agreed well with each other. But MBs and NMBs had some difference. MBs showed that the concentration of $PM_{2.5}$ was underestimated in 2014 and 2015 while overestimated in 2016 and 2017 although the biases were not very large. O_3-8h was underestimated in 2015 and overestimated in the other 3 years. NMBs for $PM_{2.5}$ in 2017 and for O_3-8h in 2014 were relatively larger. In general, most statistical indicators for different years were comparable with each other.

To evaluate the capability of RAEMS on forecasting pollution, the categorical performance was calculated using the definition referenced in [20, 38] and the results are listed in **Table 3**. Only one heavy pollution for O_3-8h (>265) occurred and therefore it was not included in the analysis. The exceedance limits were set using the criterion values for lightly, moderately, and heavily ($PM_{2.5}$ only) polluted level in the technical regulation of CNAAQS2012. The results

	R	MB	RMSE	NMB (%)	FAC2	R	MB	RMSE	NMB (%)	FAC2
All	0.74	0.1	22.3	8.1	0.89	0.80	3.5	30.6	7.2	0.95
2014	0.75	−0.7	23.9	4.4	0.89	0.80	15.8	32.0	21.3	0.96
2015	0.78	−5.6	22.5	−2.3	0.89	0.81	−6.5	30.0	−1.4	0.94
2016	0.73	1.3	22.1	13.3	0.89	0.76	2.1	32.9	5.9	0.93
2017	0.75	5.3	20.8	17.2	0.88	0.86	2.8	28.2	3.2	0.97

Table 2. Summarized statistics of forecast performance of daily $PM_{2.5}$ (left panel) and O_3-8h (right) for different forecast length (units: μg/m³ for MB and RMSE).

Exceedance limit (μg/m³)	75	115	150	160	215
Accuracy (%)	87.2	95.0	98.5	91.5	97.0
Probability of detection (%)	63.5	44.1	40.0	75.3	67.5
Missed detection rate (%)	36.5	55.9	60.0	24.7	32.5
False alarm rate (%)	43.5	60.6	68.4	42.3	52.6
Critical success index (CSI)	0.43	0.26	0.21	0.49	0.39
Bias	1.12	1.12	1.3	1.31	1.43

Table 3. Categorical performance evaluated with different exceedance limits for $PM_{2.5}$ (left panel) and O_3-8h (right).

showed that the forecast performance decreases with increased exceedance limits for both $PM_{2.5}$ and O_3-8h. The values probability of detection and critical success index decrease with higher exceedance limit, while those of missed detection rate and false alarm rate increase. The biases are relatively steady and show slight over-estimation for $PM_{2.5}$ and some for O_3-8h. An interesting result is found for accuracy that it tends to increase with higher exceedance limits. Further analysis showed that this result is ascribed to the big percentage of the records under limits.

In general, RAEMS makes good performance on forecasting the major air pollutants over eastern China region. It also provides reliable products to support and promote the work on environmental meteorology and positive effects on increasing the ability to serve the decision-making and the public.

4. Uncertainty in forecasting air quality

The previous studies also showed shortage and uncertainty in several aspects in simulating and forecasting air quality using numerical models, although great improvements were achieved. The outputs of air pollutant concentrations from numerical models are more or less different from the observations in most cases. In other words, the bias of prediction and observation is usually more than 10%. If the forecast performances well, the bias could be even less than 10% (e.g. in [20, 32, 38]). For the ratio modeled value within between half and twice of observation, good performance could be around 90% in this work, while 70–80% [38, 39] or lower [58] were more recorded. Moreover, the temporal variation of model always varies from that of observation. This can be represented in correlation coefficient or ozone peak time as one often focused issue. High correlation coefficients could be greater than 0.7 or even 0.8 (in this work and [32, 38]), usually 0.5 or 0.6 (in [20, 32, 38, 39]) or lower (in [36]). A certain percentage of forecasted ozone peak time was several hours different from observed [32, 38]. The third aspect is that model performance is generally inconsistent in space, in other words, it may perform very well over some areas but poorly over some other areas in the same simulation using the same model. This phenomenon of inconsistency existed in results of all work. The models are not as satisfied in polluted situations as in usual or clean conditions while pollution always takes more attention in many regions. For example, RAEMS did not provide enough satisfied forecast for air pollution, especially heavy pollution for Shanghai shown in former sections as well as in [20, 38] which showed unsatisfied results for high ozone in US. The performance on predicting aerosol components was worse than that on the integrated mass concentration (e.g. $PM_{2.5}$ and PM_{10}) (e.g. in [19, 20, 32]). This concerns to visibility and haze related forecast, which leads to lower capability of models in forecasting visibility and haze events.

Major components which caused the uncertainty on numerical air quality modeling and forecasting could be classified into the several following issues. First of all, emission inventories are important as they were always mentioned in many previous studies [21, 32, 38, 41]. Emissions can be classified into natural emissions and anthropogenic emissions. Natural emissions are from respiration and photosynthesis of plants, sea spray, forest fire, volcano explosion, etc. Many sources of deviation could be included in the model calculation because it's impossible for modelers to know all of the details that can influence emission. For example, it is hard to obtain the fully accurate information on the growing states and types of plants, ambient conditions such as temperature, humidity, radiance, etc. in the region and duration to be forecasted or modeled. In forecasting, it is also difficult to know exactly when, where or even whether a forest fire or volcano explosion will occur or not. There are also many kinds of uncertainties in calculating the anthropogenic emissions. The inventory is always 2 or 3 years delayed and supplies the total amount of emission for 1 month or 1 year. In most situations, the diurnal variations used in the modeling are solid in time and space and cannot describe the tempo-spatial change due to actual activities of industry, traffic, etc. Another gap is that basic monitoring data is not sufficient enough for producing anthropogenic emission inventory in chemical species and spatial resolution, and therefore many approaches are implemented in developing inventories. At the same time, inventories are also sufficient enough for modeling, e.g. the number and types of chemical species and the height of each power plant.

The second uncertainty came from model representation. While developing a model, scientists always endeavored to balance the scientific understanding and the goal of extremely "perfect" performance. But in fact, a perfect model is always idealized and being sought. The understanding of the chemical processes formatting or depleting air pollutants, the physical processes that transport or disperse air pollutants, the ambient conditions that influence chemical reactions is advancing. Forecast models usually introduce relatively mature technologies and keep them suitable for most situations. New technology is always developing to study or solve problems and be implemented into forecast model when it is validated. So, air quality models were in progress in the past and there is still some shortage or uncertainty in "current" model. Concerning RAEMS, its core model was developed several years ago and some elements were not included which were confirmed to influence the performance. For example, aerosol direct forcing in solar radiation was not considered in the model, which leads to more solar radiative flux to the air near ground and to ground surface. This deficiency results in higher near surface wind speed, PBL height and stronger vertical diffusion and thus lower primary pollutants and $PM_{2.5}$ [21, 41]. This model missed some heterogeneous uptake of sulfate under high relative humidity conditions. For example, Wang et al. [14] and Cheng et al. [25] found a new source from reactive nitrogen chemistry in aerosol water, which explained the missing of sulfate and particle matter in extreme pollution conditions in northern China region.

Bias may come from the treatments and inputs of initial and lateral boundary conditions. Usually, input data for initial and boundary conditions includes biases comparing with "real" atmosphere and is coarser than regional air quality model. More on this issue in meteorological predictions can be found in the other chapters and chemical aspects are analyzed here. Specific to RAEMS, the inputted meteorological data is 0.5 degree and much coarser than the model resolution of 6-km. The interval of 6-h may also involve bias in calculating the tendency of meteorological variables. The treatment of lateral boundary conditions in chemistry using historic mean field may make them far from reality. The missing of assimilation on both meteorological and chemical variables produced initial bias. The impact of such missing on air pollutants may exist in several hours since the model start over strong emitting regions but last for a long time over downwind regions, as the effect of chemical assimilation can be kept within 12–24 h [59]. Better initial chemical conditions are strongly needed for nowcasting of air quality.

The uncertainty in meteorological variables could be another important source. It is known that meteorological variables are drivers of CTMs. Some of them drive the processes of advection, convection, dispersion, turbulent mixing, etc. Some of them participate in chemical reactions such as vapor or decide the reactivity rate. This chapter will not focus on this for much discussion, however, this uncertainty can be found in other chapters which concern meteorology prediction. But one point we should emphasize is that the uncertainty in forecast of weak weather conditions will be paid more attention to because heavy air pollution often occurs under such conditions, although weak conditions are not so focused in meteorology for less extreme weather occurring.

5. Work in the future

To improve the performance of numerical air quality forecast, several types of work are taken into consideration in the future. As one important application of numerical meteorological prediction and the role of meteorological variables driving CTMs, introduction of better numerical forecast of weather is always one economical and effective way to improve air quality forecast. This way should be carried out indubitably if it is feasible in technology.

Update in emission inventory and its implementation in CTMs is another core action. It includes several aspects: (1) reduction of time delay; (2) increase in horizontal and vertical resolution; (3) improving the accuracy of emission inventory itself; and (4) improving the applicability in models. The former three aspects mainly require efforts of inventory community and the last one needs efforts of modelers. Specifically, one job is to improve on-line calculated emissions, such as biogenic volatile organic compounds. For example, biogenic emissions can be calculated using model meteorological variables and some inputted static data in many current CTMs (e.g. WRF-Chem and CMAQ). Better vegetation data (classification, leaf area, etc.) will benefit improvements of biogenic emissions and they can be retrieved from satellite data nearly real time. The other is to build one fast technology to adjust the emission data inputted into forecast system. The determination of indicators which may be used to adjust the emission data will be the first step and then develop a relatively fast evaluation system or technology to supply the result how the forecast performed in previous duration.

Based on the evaluation results, a fast adjustment technology is to be implemented to update emissions used in the coming forecasts. Besides the regular treatments, fast response to emergency or temporary emission control needs to be prepared based on relatively less detailed information.

To fit the extending needs, numerical air quality forecast is increasing its capability on longer predictable period, finer resolution, and better service for other interests. Long time length and fine scale is the two main aims or requirements of coming air quality prediction besides higher accuracy. Long prediction of over 1 week has been urgently needed and required by decision-making agencies during recent years. Under the strong requirements on improving air quality, environment protection agencies over eastern China often carry out or be demanded to carry out temporary emission control to reduce air pollution. This action usually needs a few days ahead of predicted pollution episode. Another important need is on macro-management of industrial production, electric power, etc. for long-term objectives such as the level of annual mean air pollutant concentrations and the level of days of pollution. It requires climate scale prediction of air quality, such as monthly or seasonally. The other aim is finer forecast in space and time. For example, tasks of air quality forecast for a specific community or a specific time point were required, which were far beyond the capability of current forecast service 3 times a day for the entire Shanghai. Many other interests, such as human health service, also need the support of numerical air quality forecast. These needs require supporting information beyond forecast results to promote their own goals.

Comparing with that in meteorological prediction, treatment and approach in initial and boundary layer conditions is rough and ongoing. Assimilation on air quality related variables or chemical assimilation is needed to improve initial conditions and forecast performance, especially in nowcasting of air quality. Of course, chemical assimilation is more difficult than meteorological assimilation due to insufficient monitoring data. Implementation of real time global forecast in boundary is another way to reduce bias from lateral input out of the model domain. This treatment will greatly benefit the forecasts near model lateral boundary and of long-term period.

Improvement in representation of CTMs such as involving the feedback and interaction between meteorological variables and air pollutants is one persistent work. This work will provide better models for numerical air quality forecast and is essential for improving model performance. But it depends on scientific understanding and technological maturity. Some nowaday jobs could focus on increasing model performance on near-surface wind, vertical diffusion of particles, aerosol species, and diurnal variation in operational forecast. We should show more desire to involve new technology into forecast system in the future.

6. Conclusion

Air pollution is focused because of its adverse effects, e.g. on human health. Numerous works were taken into action including scientific study, policy making, and emission control, etc. over eastern China due to the severe situation as one of the most polluted areas. This chapter illustrated the numerical forecast of air quality over the eastern China region, especially what has been done in Shanghai Meteorological Service.

http://dx.doi.org/10.5772/intechopen.79304

Numerical air quality forecast has become truly profiting from the achievements on air quality models and computation technology during past decades. Three-dimensional chemical transport models were the major choice in studying and predicting air quality in both global and regional scale after entering the twenty-first century. In very recent years, online approach CTMs, which calculate meteorological and chemical variables in one model, prevent from the loss of information between two meteorological outputs, and benefit involving the interaction between meteorology and air pollution. The fully online coupled WRF-Chem was chosen to develop the Regional Atmospheric Environmental Modeling System for eastern China by SMS for its good performance in modeling the air quality/pollution over this region.

The operational RAEMS was certified by China Meteorological Administration in March 2013 and has been providing numerical forecast data and products from then on. This forecasts greatly promoted the air quality prediction, air pollution warning, and decision-making service in meteorological agencies as well as environmental protection agencies. A previous detailed evaluation validated the performance on forecasting the spatial distribution and temporal variation of major air pollutants over eastern China region during the 2 years of 2014–2015 [38]. For the two most important air pollutants of $PM_{2.5}$ and O_3-8h for the city of Shanghai, RAEMS had excellent performance during 2014–2017 as analyzed in this chapter. At the same time, RAEMS showed relatively lower accuracy under polluted conditions than unpolluted conditions, and it even performed worse under heavier polluted conditions.

Further analysis showed that shortage or uncertainty in current numerical air quality forecast mainly came from four aspects of emission inventory or emission related inputs, model capability in chemical representation, biases in initial and lateral boundary layer conditions, and uncertainty in meteorological variables. These suggested ideas for improving performance of forecasts in the future. Longer predictable period and finer temporal and spatial resolution is also important goal and challenge for fitting the extending needs from application communities.

Acknowledgements

The development of RAEMS was a joint work in Shanghai Meteorological Service and collaborated with many colleagues such as Jianming Xu, Fuhai Geng, Li Peng, Ying Xie, etc. and guided by many experts e.g. Xuexi Tie, Greg Carmichael, and Georg Grell, etc. This work was sponsored by the National Key R&D Program of China (Grant nos. 2016YFC0201900 and 2016YFC0203400).

Conflict of interest

The authors declared that they have no conflicts of interest to this work.

Author details

Guangqiang Zhou*, Zhongqi Yu, Yixuan Gu and Luyu Chang

*Address all correspondence to: zhougq21@163.com

Shanghai Meteorological Service, Shanghai, P. R. China

References

[1] Donkelaar A, Martin R, Brauer M, et al. Global estimates of ambient fine particulate matter concentrations from satellite-based aerosol optical depth: Development and application. Environmental Health Perspectives. 2010;**118**(6):847-855

[2] Chan CK, Yao XH. Air pollution in megacities in China. Atmospheric Environment. 2008; **42**:1-42

[3] Geng F, Tie X, Xu J, et al. Characterizations of ozone, NOx, and VOCs measured in Shanghai, China. Atmospheric Environment. 2008;**42**:6873-6883

[4] Zhang X, Wang YQ, Niu T, et al. Atmospheric aerosol compositions in China: Spatial/ temporal variability, chemical signature, regional haze distribution and comparisons with global aerosols. Atmospheric Chemistry and Physics. 2012;**11**:26571-26615

[5] Stocker TF, Qin D, Plattner GK, et al., editors. IPCC, 2013: Climate change 2013: The physical science basis. In: Contribution of Working Group I to the Fifth Assessment Report of the Intergovernmental Panel on Climate Change. Cambridge, United Kingdom and New York, NY, USA: Cambridge University Press. 1535 pp

[6] Zhang M, Song Y, Cai X, et al. Economic assessment of the health effects related to particulate matter pollution in 111 Chinese cities by using economic burden of disease analysis. Journal of Environmental Management. 2008;**88**:947-954

[7] Lim SS, Theo V, Flaxman AD, et al. A comparative risk assessment of burden of disease and injury attributable to 67 risk factors and risk factor clusters in 21 regions, 1990-2010: A systematic analysis for the global burden of disease study 2010. Lancet. 2012;**380**(9859): 2224-2260

[8] Chen R, Chen C, Tan J, et al. Ambient air pollution and hospital admission in Shanghai, China. Journal of Hazardous Materials. 2010;**181**(1-3):234-240

[9] Geng F, Hua J, Mu Z, et al. Differentiating the associations of black carbon and fine particle with daily mortality in a Chinese city. Environmental Research. 2013;**120**:27-32

[10] Huang W, Tan J, Kan H, et al. Visibility, air quality and daily mortality in Shanghai, China. Science of the Total Environment. 2009;**407**(10):3295-3300

[11] Shang Y, Sun Z, Cao J, et al. Systematic review of Chinese studies of short-term exposure to air pollution and daily mortality. Environment International. 2013;**54**:100-111

[12] Cao C, Jiang W, Wang B, et al. Inhalable microorganisms in Beijing's PM2.5 and PM10 pollutants during a severe smog event. Environmental Science & Technology. 2014;**48**: 1499-1507

[13] Guo S, Hu M, Zamora ML, et al. Elucidating severe urban haze formation in China. Proceedings of the National Academy of Sciences. 2014;**111**:17373-17378

[14] Wang Y, Zhang Q, Jiang J, et al. Enhanced sulfate formation during China's severe winter haze episode in Jan 2013 missing from current models. Journal of Geophysical Research-Atmospheres. 2014;**119**:10425-10440

[15] Zhang YM, Zhang XY, Sun JY, et al. Chemical composition and mass size distribution of PM_1 at an elevated site in central east China. Atmospheric Chemistry and Physics. 2014;**14**: 12237-12249

[16] Ministry of Environmental Protection of the People's Republic of China and General Administration of Quality Supervision, inspection and quarantine of the People's Republic of China: National ambient air quality standards (CNAAQS2012, GB 3095-2012), 2012.

[17] Action projects on prevention and control of air pollution, 2013. Available from: http://www.gov.cn/zwgk/2013-09/12/content_2486773.htm [Accessed: 2018-05-12]

[18] Tie X, Geng F, Peng L, et al. Measurement and modeling of O_3 variability in Shanghai, China: Application of the WRF-Chem model. Atmospheric Environment. 2009;**43**:4289-4302

[19] Zhou G, Yang F, Geng F, et al. Measuring and modeling aerosol: Relationship with haze events in Shanghai, China. Aerosol and Air Quality Research. 2014;**14**:783-792

[20] Grell GA, Peckham SE, Schmitz R, et al. Fully coupled 'online' chemistry in the WRF model. Atmospheric Environment. 2005;**39**:6957-6976

[21] Wang ZF, Li J, Wang Z, et al. Modeling study of regional severe hazes over mid-eastern China in January 2013 and its implications on pollution prevention and control. Science China Earth Sciences. 2014;**57**(1):3-13

[22] Byun DW, Schere KL. Review of the governing equations, computational algorithms, and other components of the Models-3 community multiscale air quality (CMAQ) modeling system. Applied Mechanics Reviews. 2006;**59**:51-77

[23] Zheng B, Zhang Q, Zhang Y, et al. Heterogeneous chemistry: A mechanism missing in current models to explain secondary inorganic aerosol formation during the January 2013 haze episode in North China. Atmospheric Chemistry and Physics. 2015;**15**:2031-2049

[24] Zheng G, Duan F, Su H, et al. Exploring the severe winter haze in Beijing: The impact of synoptic weather, regional transport and heterogeneous reactions. Atmospheric Chemistry and Physics. 2015;**15**:2969-2983

[25] Cheng Y, Zheng G, Wei C, et al. Reactive nitrogen chemistry in aerosol water as a source of sulfate during haze events in China. Science Advances. 2016;**2**:e1601530

[26] Byun DW, Ching JKS, Novak J, Young J. Development and implementation of the EPA's models-3 initial operating version: Community multi-scale air quality (CMAQ) model. In: Gryning SE, Chaumerliac N, editors. Air Pollution Modeling and its Application XII. B oston: Springer; 1998. pp. 357-368. DOI: 10.1007/978-1-4757-9128-0_37

[27] Chang JS, Binkowski FS, Seaman NL, et al. The regional acid deposition model and engineering model. In: State-of-Science/Technology, Report 4, National Acid Precipitation Assessment Program. Washington, DC; 1989

[28] Horowitz LW, Walters S, Mauzerall D, et al. A global simulation of tropospheric ozone and related tracers: Description and evaluation of MOZART, version 2. Journal of Geophysical Research. 2003;**108**(D24):4784

[29] Emmons L, Walters S, Hess P, et al. Description and evaluation of the model for ozone and related chemical tracers, version 4 (MOZART-4). Geoscientific Model Development. 2010;**3**:43-67

[30] Gu Y, Liao H, Bian J. Summertime nitrate aerosol in the upper troposphere and lower stratosphere over the Tibetan Plateau and the South Asian summer monsoon region. Atmospheric Chemistry and Physics. 2016;**16**:6641-6663. DOI: 10.5194/acp-16-6641-2016

[31] Gu Y, Liao H, Xu J, Zhou G. The chemical effects on the summertime ozone in the upper troposphere and lower stratosphere over the Tibetan Plateau and the South Asian monsoon region. Meteorology and Atmospheric Physics. 2018;**D17**:1-11

[32] Chen J, Vaughan J, Avise J, et al. Enhancement and evaluation of the AIRPACT ozone and $PM_{2.5}$ forecast system for the Pacific Northwest. Journal of Geophysical Research-Atmospheres. 2008;**113**(D14):762-770

[33] Foley KM, Roselle SJ, Appel KW, Bhave PV. Incremental testing of the community multiscale air quality (CMAQ) modeling system version 4.7. Geoscientific Model Development. 2010;**3**(1):205-226

[34] Tie X, Brasseur G, Ying Z. Impact of model resolution on chemical ozone formation in Mexico City: Application of the WRF-Chem model. Atmospheric Chemistry and Physics. 2010;**10**:8983-8995. DOI: 10.5194/acp-10-8983-2010

[35] Zhou CH, Gong SL, Zhang XY, Wang YQ. Development and evaluation of an operational SDS forecasting system for East Asia: CUACE/Dust. Atmospheric Chemistry and Physics. 2008;**8**(4):787-798

[36] Zhou CH, Gong SL, Zhang XY, et al. Towards the improvements of simulating the chemical and optical properties of Chinese aerosols using an online coupled model – CUACE/Aero. Tellus B. 2012;**64**(1):91-102

[37] Gao M, Carmichael GR, Wang Y, et al. Modeling study of the 2010 regional haze event in the North China plain. Atmospheric Chemistry and Physics. 2016;**15**(16):22781-22822

[38] Zhou G, Xu J, Xie Y, et al. Numerical air quality forecasting over Eastern China: An operational application of WRF-Chem. Atmospheric Environment. 2017;**153**:94-108. DOI: 10.1016/j.atmosenv.2017.01.020

[39] Wu J, Wang Z, Wang Q, et al. Development of an on-line source-tagged model for sulfate, nitrate and ammonium: A modeling study for highly polluted periods in Shanghai, China. Environmental Pollution. 2017;**2017**(221):168-179

[40] Li L, An JY, Shi YY, et al. Source apportionment of surface ozone in the Yangtze River Delta, China in the summer of 2013. Atmospheric Environment. 2016;**144**:194-207. DOI: 10.1016/j.atmosenv.2016.08.076

[41] Zhang B, Wang Y, Hao J. Simulating aerosol–radiation–cloud feedbacks on meteorology and air quality over eastern China under severe haze conditions in winter. Atmospheric Chemistry and Physics. 2015;**15**:2387-2404

[42] Fast JD, Gustafson WI, Easter RC, et al. Evolution of ozone, particulates, and aerosol direct radiative forcing in the vicinity of Houston using a fully coupled meteorology-chemistry-aerosol model. Journal of Geophysical Research-Atmospheres. 2006;**111**(D21):5173-5182

[43] Operational and Experimental Air Quality Forecasts Using WRF-Chem. 2009. Available from: https://ruc.noaa.gov/wrf/wrf-chem/Real_time_forecasts.htm [Accessed: 2018-05-12]

[44] Tie X, Madronich S, Li GH, et al. Characterizations of chemical oxidants in Mexico City: A regional chemical/dynamical model (WRF-Chem) study. Atmospheric Environment. 2007;**41**:1989-2008

[45] Fountoukis C, Nenes A. ISORROPIA II: A computationally efficient aerosol thermodynamic equilibrium model for K^+, Ca^{2+}, Mg^{2+}, NH_4^+, Na^+, SO_4^{2-}, NO_3^-, Cl^-, H_2O aerosols. Atmospheric Chemistry and Physics. 2007;**7**:4639-4659

[46] Schell B, Ackermann I, Hass H, et al. Modeling the formation of secondary organic aerosol within a comprehensive air quality model system. Journal of Geophysical Research. 2001;**106**:28275-28293

[47] Madronich S, Flocke S. The role of solar radiation in atmospheric chemistry. In: Boule P, editor. Handbook of Environmental Chemistry. Heidelberg: Springer; 1999. pp. 1-26

[48] Tie X, Madronich S, Walters S, et al. Effect of clouds on photolysis and oxidants in the troposphere. Journal of Geophysical Research. 2003;**108**:4642

[49] Guenther A, Karl T, Harley P, et al. Estimates of global terrestrial isoprene emissions using MEGAN (model of emissions of gases and aerosols from nature). Atmospheric Chemistry and Physics. 2006;**6**:3181-3210

[50] Guenther A, Zimmerman P, Wildermuth M. Natural volatile organic compound emission rate estimates for US woodland landscapes. Atmospheric Environment. 1994;**28**:1197-1210

[51] Li M, Zhang Q, Streets D, et al. Mapping Asian anthropogenic emissions of non-methane volatile organic compounds to multiple chemical mechanisms. Atmospheric Chemistry and Physics. 2014;**14**:5617-5638

[52] Liu F, Zhang Q, Tong D, et al. High-resolution inventory of technologies, activities, and emissions of coal-fired power plants in China from 1990 to 2010. Atmospheric Chemistry and Physics. 2015;**15**(13):18787-18837

[53] Zhang Q, Streets DG, Carmichael GR, et al. Asian emissions in 2006 for the NASA INTEX-B mission. Atmospheric Chemistry and Physics. 2009;**9**:5131-5153

[54] Sharing Platform of Environmental Meteorological information for East China, 2013. Available from: http://222.66.83.20:801 [Accessed: 2018-05-12]

[55] Zhang YW, Zhang XY, Zhang YM, et al. Significant concentration changes of chemical components of PM_1 in the Yangtze River Delta area of China and the implications for the formation mechanism of heavy haze-fog pollution. Science of the Total Environment. 2015;**538**:7-15

[56] Liang L, Engling G, Zhang X, et al. Chemical characteristics of PM2.5, during summer at a background site of the Yangtze River Delta in China. Atmospheric Research. 2017;**198**: 163-172

[57] Derwent D, Fraser A, Abbott J, et al. Evaluating the Performance of Air Quality Models. London: Department for Environment, Food & Rural Affairs of UK; 2011. p. 73

[58] Prank M, Sofiev M, Tsyro S, et al. Evaluation of the performance of four chemical transport models in predicting the aerosol chemical composition in Europe in 2005. Atmospheric Chemistry and Physics. 2016;**16**:1-42

[59] Wu J, Xu J, Mariusz M, et al. Modeling study of a severe aerosol pollution event on December 2013 in Shanghai China: Application of chemical data assimilation. Particuology. 2015;**20**:41-51

Chapter 9

Numerical Simulation of the Effects of Increasing Urban Albedo on Air Temperatures and Quality over Madrid City (Spain) by Coupled WRF/CMAQ Atmospheric Chemistry Model

Pablo Campra

Additional information is available at the end of the chapter

http://dx.doi.org/10.5772/intechopen.80473

Abstract

Meteorological and photochemical impacts of increasing urban albedo or reflectance over Madrid city have been simulated using a mesoscale climatic model (WRF) coupled to an air quality modeling system (AEMM/CMAQ). We have evaluated the influence over the concentration of the main pollutants of two different interventions with increasing levels of albedo enhancement over all urban categories: a low albedo or cool roofs scenario (Alb1), where only roof albedo was modified (+0.35), and a high albedo scenario (Alb2), increasing both roof albedo (+0.35) and pavement albedo (+0.15). Simulations were run for two periods of 72 h, representative of summer and winter conditions. In both scenarios, surface air temperatures were cooled, with averaged midday reductions at the urban area of −0.2 (−0.5)°C for winter (summer) for Alb1 and −0.3 (−0.7)°C for winter (summer) for Alb2. Peak summer midday cooling at city center was −1.4°C and −1.6 °C for Alb1 and Alb2, respectively. Pollutant concentrations were modified, with reductions in O_3 levels, higher in summer, and increases in NO_2 levels, bigger in winter period. Slight increases were also observed in winter for SO_2 and particulate matter ($PM_{2.5}$ and PM_{10}) in both scenarios.

Keywords: WRF, BEM, CMAQ, urban albedo, air quality, Madrid city, cool roofs

1. Introduction

Adaptation to climate change by increasing the reflectance of human settlements has been proposed as a simple and cost-effective geo-engineering strategy to offset the rise of temperatures associated with global warming at local and regional scales [1]. The use of higher albedo roofs and/or pavements (cool roofs and pavements) has shown effective surface air cooling in simulation experiments over many cities in the world [2]. Due to lower surface temperatures, an improvement in air quality can be obtained by slowing temperature-dependent photochemical reaction rates of formation of secondary pollutants, such as ozone, and reducing biogenic hydrocarbon emissions. Additional indirect benefits are linked to lower energy demand for summer cooling of buildings and its associated emissions in power plants [3]. On the contrary, due to the depression of the planetary boundary layer level (PBL) height caused by cooler temperatures, and to possible changes in local wind patterns, reduced mixing and dilution of pollutants can raise their levels by accumulation in some areas. Ground-level ozone (O_3), particulate matter (PM), nitrogen oxides (NO_x), and sulfur oxides (SO_2) are most health-concerned pollutants, and their urban concentration levels can be affected by surface modification. O_3 is a secondary pollutant resulting from the reaction between NO_x oxides and volatile organic compounds (VOCs) in the presence of sunlight. Higher O_3 concentration levels are directly related to warming in urban heat islands, reaching peak levels in summertime [4].

Most numerical simulations of the impact of albedo increase on pollutants have been developed over US cities, in general with different urban fabrics than in Europe, where compact mid-rise urban categories occupy most of the city centers. In pioneering mesoscale numerical simulations [5], extreme surface albedo enhancement resulted in ozone reductions in California. Simulating a more feasible albedo increase over southern US cities with high insolation levels, [6] obtained ozone reductions linked to cooler summer ambient air temperatures. However, only Sacramento showed significant peak ozone-level reductions due to a wider urban surface (25,000 ha). Applying comparable albedo increase levels, it has been simulated significant air quality benefits over California [7], suggesting that there is a maximum albedo increase implementation threshold above which no further benefits are obtained that should be determined for every urban case. There are scarce mesoscale simulation experiments to investigate the impact of albedo enhancement on pollutants other than ozone. In another simulation the effect on air quality during a heat wave episode of albedo increase at high latitudes (Montreal, Canada) was reported [8], with no significant effect on ozone levels, and slight reductions in $PM_{2.5}$ levels (2 ppb), associated with a decrease in PBL height, that counteracted the cooling impact on ozone formation. Applying an extreme albedo increase over Stuttgart (Germany), a peak urban temperature cooling down to −1.7°C and decreases in mean ozone concentration were reported [9]. However, secondary undesirable effects were an increase of primary pollutants (NO_x and CO) and an increase in peak ozone concentration due to a higher intensity of reflected UV shortwave radiation. To date, the impact of urban albedo enhancement on temperatures and air quality has never been assessed over Spanish urban areas by numerical modeling. Spanish cities may give key information for this research, due to the high levels of annual and summer insolation and to the hot summer Mediterranean climate in most of the country, with high annual number of clear skies that maximize the thermal and energy-saving potential benefits of changes in solar reflectivity [10]. On the other hand, an undetermined minimum critical intervention surface is also needed to obtain significant modifications in local atmospheric variables by land cover changes [6]. Madrid city is the biggest urban area in Spain, with broadly five million inhabitants in its metropolitan area, and the fourth most populated city in Europe. Emissions of air pollutants in Madrid are mostly originated from anthropogenic sources, with the traffic sector as the main contribution activity to the emissions of the whole region. In the last years, the Regional Government of Madrid has developed an ambitious action plan to improve the air quality for the period 2013–2020, called *Plan Azul+*. A WRF mesoscale simulation [11] showed that the *Plan Azul+* measurements were effective in the reduction of NO_2 levels over urban areas with high traffic influence. However, this simulation showed slight increases in ozone concentration (1–2%) in areas where typically ozone levels were low, and mitigation measures did not cause remarkable reductions in the rest of pollutants selected in the plan. Thus, prior to the establishment of recommendations for policymakers to include albedo enhancement in urban planning, the balance between potential climatic and air quality benefits and disturbances of widespread cooling the urban air must be assessed. Here, we have used a meteorological model (WRF), an emission model (AEMM), and a photochemical model (CMAQ) to assess the impact on meteorology and air quality of widespread urban albedo increase at two feasible levels of implementation: cool roofs (Alb1) and Alb2 (cool roofs + cool pavements). Changes in surface air temperatures and main pollutants are given for two 72-h period representative of summer and winter seasons.

2. Materials and methods

Numerical simulations of urban surface modification over Madrid city have been designed to test the impact of two increasing surface albedo scenarios, conducted by coupling WRF/AEMM/CMAQ. Urban layer was simulated by an urban energy model (BEM) coupled with an urban canopy model for simulations [12]. We have considered up to 10 urban categories. Air quality analysis has been focused over the main pollutants of health concern, namely O_3, NO_2, SO_2, CO, $PM_{2.5}$, and PM_{10}.

2.1. Study area

Surface modification was simulated over the urban land cover of Madrid city, which is located in the center of the Iberian Peninsula. Its geographical position and topography determine a temperate continental Mediterranean climate with cold humid winters, with temperatures usually below 0°C, and warm dry summer, with temperatures above 30°C, frequently reaching peak values over 40°C, and high nocturnal temperatures[13, 14].

2.2. Simulation domains

In **Figure 1** we show nested modeling domains over the city of Madrid. Modeling is built over a mother domain (d01) with 27 km spatial resolution, centered at 40.383° N 3.717°W, and a domain size of 2727 × 2727 km^2. This domain is intended to capture synoptic features and general circulation patterns. The first nested domain (d02), with a spatial resolution of 9 km, covers a domain size of 1575 × 1413 km^2. The third domain (d03) with 3 km of spatial resolution has a domain size of 660 × 561 km^2. The fourth domain covers the province of Madrid

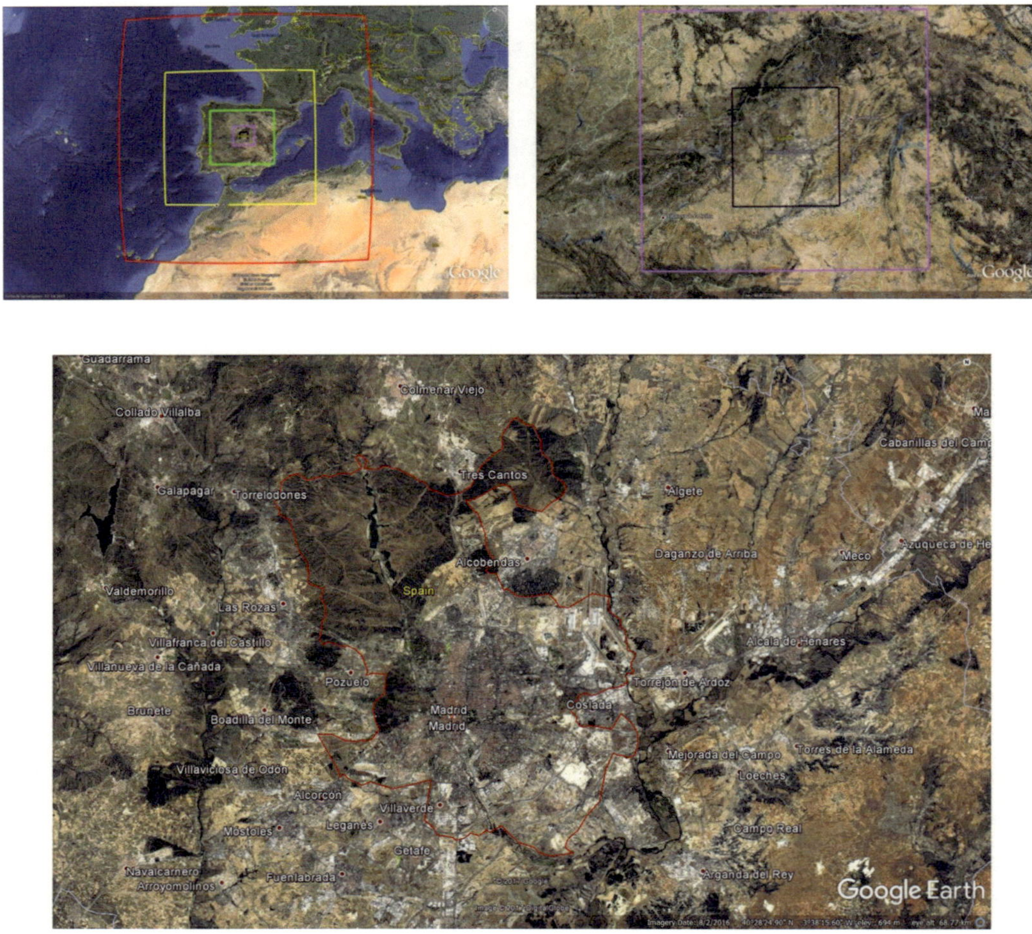

Figure 1. Up: Modelling domains for simulations: d01, d02, d03, d04 and d05 (left), and d04 and d05 (right). Down: Madrid local municipality (red line) and surrounding metropolitan area [Images generated using Google Earth].

and nearest provinces, with an extension of 217 × 199 km^2 and grid resolution of 1 km^2. The innermost fifth domain encloses Madrid metropolitan area, covering 80.3 × 90.3 km^2, and grid resolution 333 m.

2.3. Modeling approach

The air quality modeling system used to evaluate albedo scenarios was composed by a coupled WRF/AEMM/CMAQ model. To configure it, we have followed the guidelines indicated in the Guide on the use of models for the European Air Quality Directive [15]. Emission and photochemical modeling configuration used here have been previously validated elsewhere [11], using a numerical deterministic evaluation during the development of the *Plan Azul+*, considering the Maximum Relative Directive Error [15] referred in the European Directive EC/2008/50. Meteorological simulations have been performed using the

Weather Research and Forecasting-Advanced Research WRF (WRF-ARW) version 3.2 [16], developed by the National Center of Atmospheric Research (NCAR). Urban categories have been adapted from The World Urban Database and Access Portal Tools [17, 18]. URBPARM.TBL was an adapted file for Madrid city according to our knowledge of urban morphology. Meteorology-Chemistry Interface Processor (MCIP) version 4.3 was used to prepare WRF output to the photochemical model. The annual anthropogenic emissions inventory of the Regional Government of Madrid has been used (version 2010). This inventory has a horizontal resolution of 1 × 1 km^2 and includes emissions classified by Selected Nomenclature for Air Pollution (SNAP) sectors. We have used an Air Emission Model (AEMM) [11, 19] to adapt emissions to domains d04 and d05, using monthly and weekly profiles from the Unified EMEP model and vertical profiles from [20]. Emissions have been adapted under the requirements of the chemical modules considered in the photochemical model CMAQ. We have considered only anthropogenic emissions to avoid the influence of albedo modifications over the natural emissions, since the parameterizations that define them depend on meteorological conditions. This assumption is valid considering that the urban metropolitan area of Madrid is strongly dominated by anthropogenic local sources, being the natural emissions not very important, and only provide a remarkable contribution in areas far away of the city of Madrid [11]. To simulate the physical and chemical processes into the atmosphere, the US Environmental Protection Agency models-3/CMAQ model has been used [21]. Here, we have used CMAQ v5.0.1, considering CB-5 chemical mechanism and associated EBI solver [22] and AERO5 aerosol module [23]. Initial and boundary conditions for d04 domain are used from inner profiles and for d05 conditions are provided by the results of simulation of d04 domain. Coupled WRF/AEMM/CMAQ has been validated over Madrid city [11]. Another assessment of coupled WRF/CMAQ over Madrid city was reported [24], for an annual period and 1 km resolution, including a comparison of meteorological and air quality observations between WRF bulk urban canopy parameterization (UCP) (used here) and an alternative building energy model (BEP). As temperature predictions were not improved by WRF-BEP and the differences on wind direction and PBL height were not remarkable, we decided to use bulk UCP for computing time savings. Simulations have been executed over a computing cluster owned by Meteosim SL (Spain) and formed by 28 nodes and 308 cores.

2.4. Simulation results

Meteorological, emissions, and photochemical simulations have been conducted for two 72 h periods representative of both summer and winter of the year 2008: the period between June 30, 2008 and July 2, 2008 (hereinafter referred to as summer), and the period between January 1, 2008 and January 3, 2008 (hereinafter referred to as winter). The previous 24 h were taken as spin-up time to minimize the effects of initial conditions. A total amount of six simulations have been done for each period and three increasing albedo scenarios, defined from feasible levels of intervention obtained from literature [2, 25] as: (a) default scenario: using default value of albedo for all urban categories; (b) cool roofs scenario (Alb1): increasing from 0.20 to 0.55 the roof surface albedo for all urban categories; (c) cool roofs + pavement scenario (Alb2), with the same roof albedo increase as Alb1 plus an increase in pavement surface albedo from 0.15 to 0.30 for all urban categories. Spatial distributions and changes in pollutants levels at the innermost domain are reported, given as recommended time-weighted exposure parameters [26].

3. Results and discussion

3.1. Albedo changes

Averaged changes in surface albedo at the innermost domain resulting from the modification of roofs and pavements albedo over urban land cover are shown in **Figure 2a** and **b**. Maximum albedo change occurred as expected at the most compact area of town center, with +0.2 for both scenarios. No significant difference observed in this peak albedo increase between both intervention levels (data not shown for Alb2), was due to low ratio pavements/roofs at the dense urban structure of the urban category where they occur, tagged as compact mid-rise in the WUDAPT. On the contrary, surface albedo change differences were observed between scenarios at the open mid-rise category (enclosing most of the rest of the urban area inside the M40 highway belt), with +0.06 for Alb1 and +0.1 for Alb2, respectively. These two categories of residential use encompass the major proportion of urban potential intervention areas, along with the southeast belt of industrial-commercial use, where albedo increased +0.14 and +0.18, in Alb1 and Alb2, respectively.

3.2. Temperature changes

The spatial distribution of temperature changes was dependent on both the distribution and the level of albedo change. In **Figure 2c–f**, 72-h averaged changes at 12 h UTC are given. City center (compact mid-rise urban category) showed the most intense cooling in all cases, reaching the highest midday cooling intensity in summer of −1.4 and −1.6°C, for Alb1 and Alb2, respectively (**Figure 2d** and **f**). Little temperature change was observed in non-urbanized areas with small surface modification. In winter, much lower but significant cooling occurred, with maximum levels of −0.4 and −0.5°C, for Alb1 and Alb2, respectively, at the city center as well (**Figure 2c** and **e**). Thus, albedo enhancement was much more efficient in cooling air surface temperatures during summer periods, due to higher solar incidence angle. The spread of cooler air from city center toward the NE was due to predominant SW winds during the summer period, with averaging speed of 9.7 m s^{-1} (data not shown).

3.3. Changes in pollutant levels

After albedo enhancement, changes in pollutants were characterized by a decrease in O_3 in both periods, but higher in summer, and an increase in NO_2 in both periods. Averaged values for every 72-h period and scenarios are given in **Figure 3**. Spatial distributions of changes for ozone are given in **Figure 4**, in both scenarios and at the innermost domain. For the rest of pollutants, Alb1 scenario changes are given in **Figures 5** and **6**. Slight increases in PM_x and in SO_2 occurred in winter with negligible changes in summer. Little changes were observed in CO levels.

When cool pavements were added to cool roofs (Alb2), differences in distribution of pollutants other than ozone were not remarkable (Alb2 changes in **Figure 3**). Areas of major changes in pollutants after albedo increase extended through city center and NE rural areas in summer. This NE spread during the summer period was again associated due to dominant

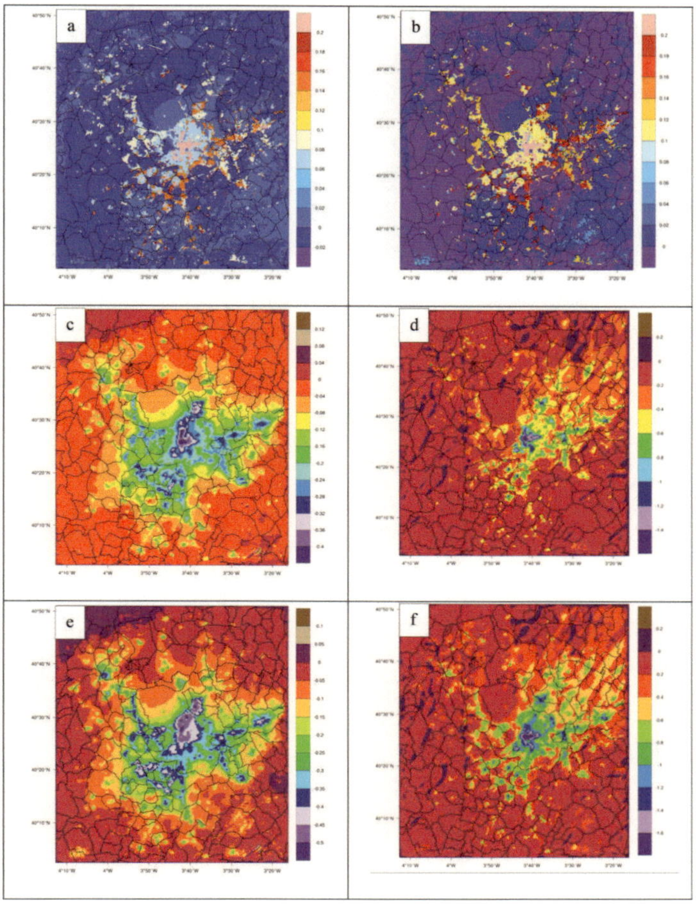

Figure 2. Spatial distribution of average albedo (a, b), and midday (12 UTC) temperature differences (°C) between default and both Alb1 (c, d) and Alb2 (e, f) scenarios, and for the winter (a, c, e) and summer periods (b, d, f) at 12 UTC. Black lines limit municipalities, the biggest and innermost boundary is Madrid city (605.77 km^2).

SW winds in those days (data not shown). Highest reduction in O_3 levels occurred during the summer period (**Figure 4b** and **d**), when more intense cooling occurred as well. Eight-hour maximum reductions of around −4 µg m^{-3} were reached at the city center. These results show that widespread cool roofs deployment over Madrid would benefit ozone levels at the city center, with additional reductions upwind depending on meteorological conditions. If additional cool pavements were implemented (Alb2), ozone reduction would extend further across most of the city, though reductions below the −4 µg m^{-3} threshold were not observed in this scenario. In winter period both scenarios show scarce benefits for ozone reduction, as expected from limited surface air temperature cooling and lower rates of ozone formation in default scenario (data not shown). Our ranges of O_3 reduction are in accordance with similar mesoscale experiments of albedo modification [7, 9] with no local increases detected in our case, as other simulation studies have reported [3]. After albedo modification, NO_2 levels increased, mainly at the city center and at Barajas airport in winter (**Figure 5a** and **b**) only

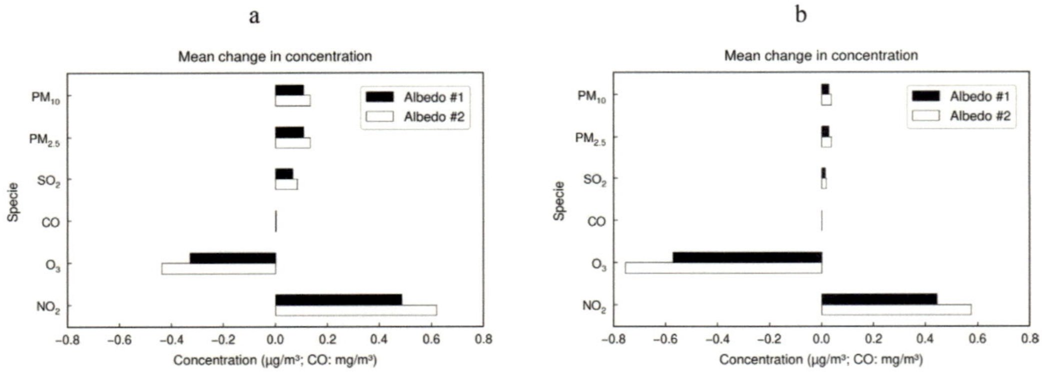

Figure 3. Mean change in pollutants concentration after albedo increase averaged for the winter (a) and the summer period (b).

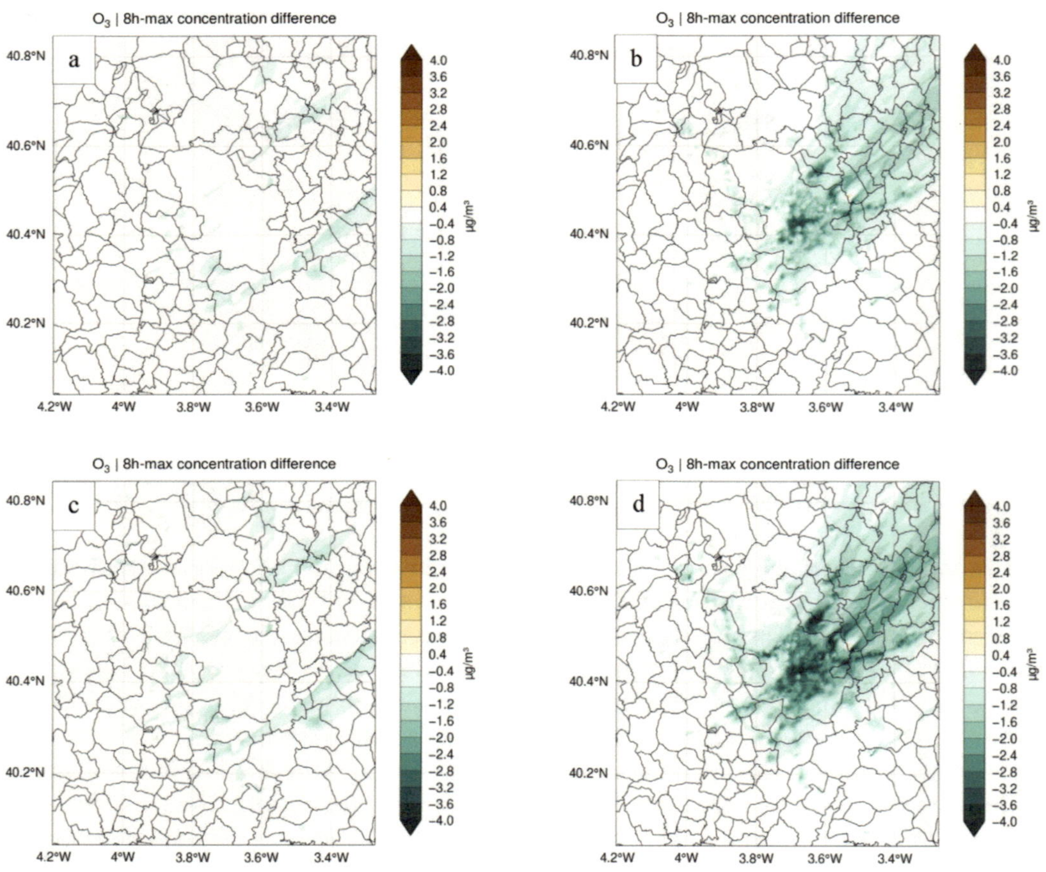

Figure 4. Spatial distribution of changes in O_3 maximum 8 h levels between default and Alb1 (a, b) and Alb2 scenarios (c, d) and for winter (a, c) and summer (b, d) ($\mu g\ m^{-3}$).

http://dx.doi.org/10.5772/intechopen.80473

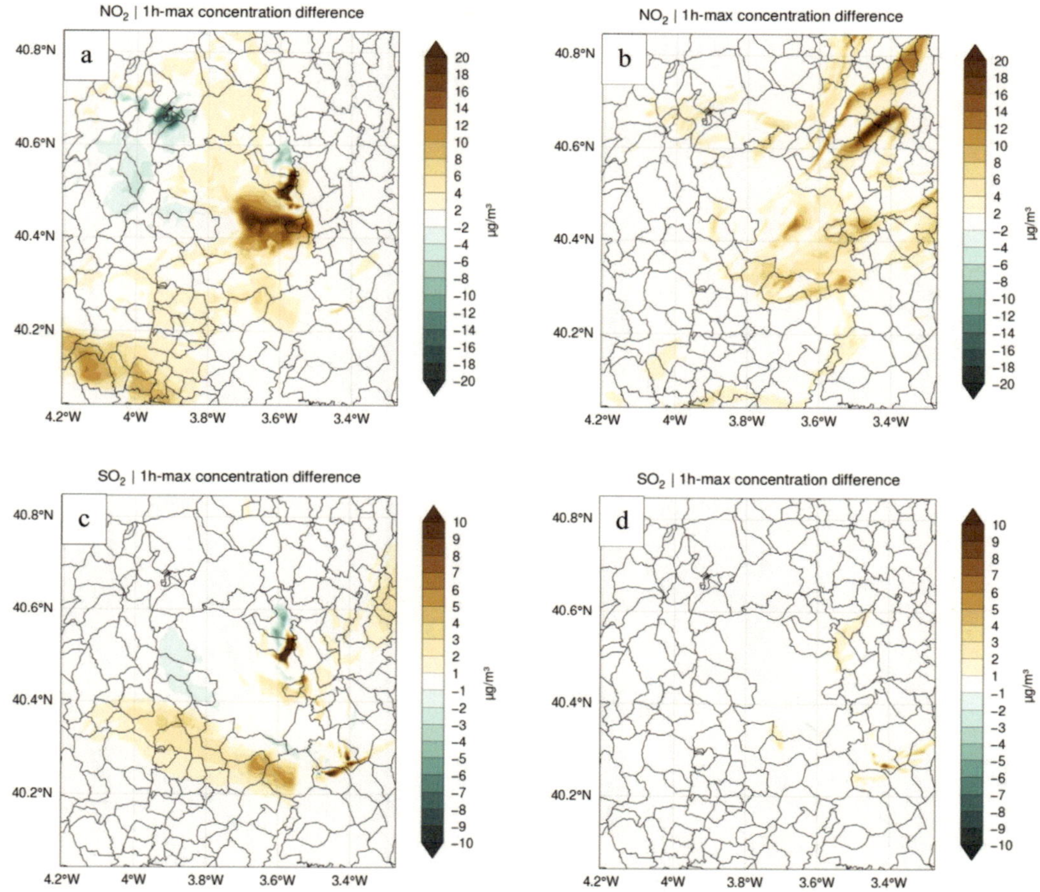

Figure 5. Spatial distribution of changes in NO_2 (a, b) and SO_2 (c, d) maximum 1 h levels between default and Alb1 scenario for winter (a, c) and summer (b, d) (µg m^{-3}).

Alb1 data are shown. Winter reductions were observed at some highly populated areas NW of Madrid city. At both periods, peak increases in 1 h—maximum concentrations reached up to +20 µg m^{-3}. At the city center, summer increases were below winter changes, though peaks of +20 µg m^{-3} were reached at the pollutants spreading area NW of the city due to wind conditions. However, and contrary to O_3 changes, spatially averaged changes in NO_2 were very similar at both periods (**Figure 5**).

Increases in SO_2 occurred in winter at the highly populated municipalities SW of the city and around the airport and NE corridor, with peak 1 h maximum differences of 10 µg m^{-3} (**Figure 5c** and **d**). Decreased levels were observed in some municipalities west of the city. No significant changes occurred for summer in SO_2 levels in Madrid city, with slight increases NE of the airport and SE of the municipality. Spatial changes between scenarios were not remarkable for the rest of pollutants (data not shown).

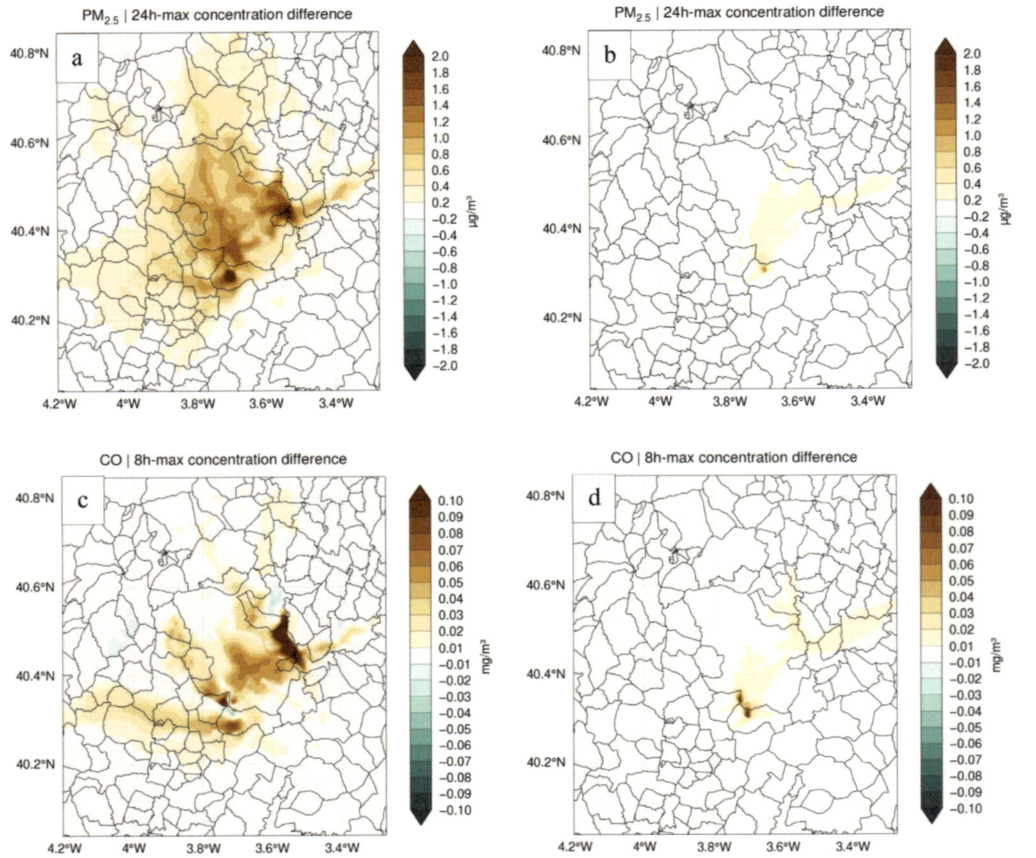

Figure 6. Spatial distribution of changes in $PM_{2.5}$ daily value (a, b) and CO 8-h maximum levels (c, d) between default and Alb1 scenarios for winter (a, c) and summer (b, d) scenarios (μg m^{-3}).

Slight differences were observed in the distribution of changes between $PM_{2.5}$ (**Figure 6a** and **b**). Increases in $PM_{2.5}$ were much higher in winter. Peak daily values increased above 2 μg m^{-3} in both scenarios, located around Barajas airport and south of the city. In summer on the contrary, little changes in $PM_{2.5}$ were observed for both scenarios, with small increases in the city center below 0.4 μg m^{-3}. For CO, small increases were detected in city center (**Figure 6c** and **d**), with maximum +0.1 mg m^{-3} also around the airport and SW of the city.

According to our simulations, it was clear that albedo enhancement caused citywide air cooling with associated pollutant changes directly linked to temperature reduction. Our model shows that cool roofs and pavements reduce outdoor temperatures slowing reaction rates of ozone formation [3]. However, observed increases in the levels of other pollutants were caused by depression of PBL height associated with cooler air, limiting pollutant dispersion, and vertical mixing. Spatial distribution matches with cooling and shows peak change levels below 100 m (**Figure 7**), in areas where pollutants show higher increments. In summer, PBL height can fall to 90 m, but lower emissions and meteorological conditions generated lower increments. O_3 summer reductions by slower formation rates must probably be partly offset by PBL fall. In terms of averaged changes from maximum levels

http://dx.doi.org/10.5772/intechopen.80473

at default scenario inside the limits of Madrid city after implementing the highest level of surface modification (Alb2 scenario) were approximately summer reductions in O_3 around to −4.4% and winter increments around +16% for NO_2, +10% for SO_2 and $PM_{2.5}$, and +5% of CO. If only cool roofs were implemented (Alb1), these maximum thresholds would be the same for all pollutants but with a lower spatial reduction at the urban center for O_3 (data not shown)

Thus, our results confirm that citywide cool roofs deployment is a feasible effective measure to reduce summer air temperatures and control the ozone pollution over Madrid metropolitan area. Further benefits can be obtained extending albedo enhancement with cool pavements. In consequence, this surface modification strategy should be considered along with other actions in air quality plans over Madrid. According to numeric simulations of the application of measures proposed in *Plan Azul+*, carried out with a similar modeling approach [11], 2% O_3 increases would be expected over Madrid region and up to +6 μg m^{-3} (1 h-maximum) at the center of town. For the rest of pollutants, the simulated effect of the *Plan* predicted reductions in NO_2 up to 11 μg m^{-3} but only slight global reductions below 5% of CO, PM_{10}, $PM_{2.5}$, and SO_2. However, our simulation shows undesired impacts of albedo enhancement that need further research before a real implementation was to be considered, as in winter cool roofs might partly offset the reductions predicted in the plan for pollutants other than ozone. Anyhow, and given the nonlinear dynamics of the atmospheric processes, it would be advisable to make new simulations for longer representative seasonal periods, combining measures of *Plan Azul+* with increasing levels of intervention of surface albedo change, to determine the balance between benefits and disadvantages on air quality of a global action plan. A key outcome of our simulation is that the increment in pollutants other than ozone occurs mainly in winter period, where ozone formation rates are low and the concentration of other pollutants is higher. On the contrary, in our summer simulation, albedo caused limited increases in these pollutants, along with a maximized ozone reduction. According to our results, an ideal implementation of cool roofs and/or pavements would be a seasonally changing system of increased reflectance only on warm periods, with little or no albedo change for colder months. Furthermore, such a system would avoid winter penalty due to increased

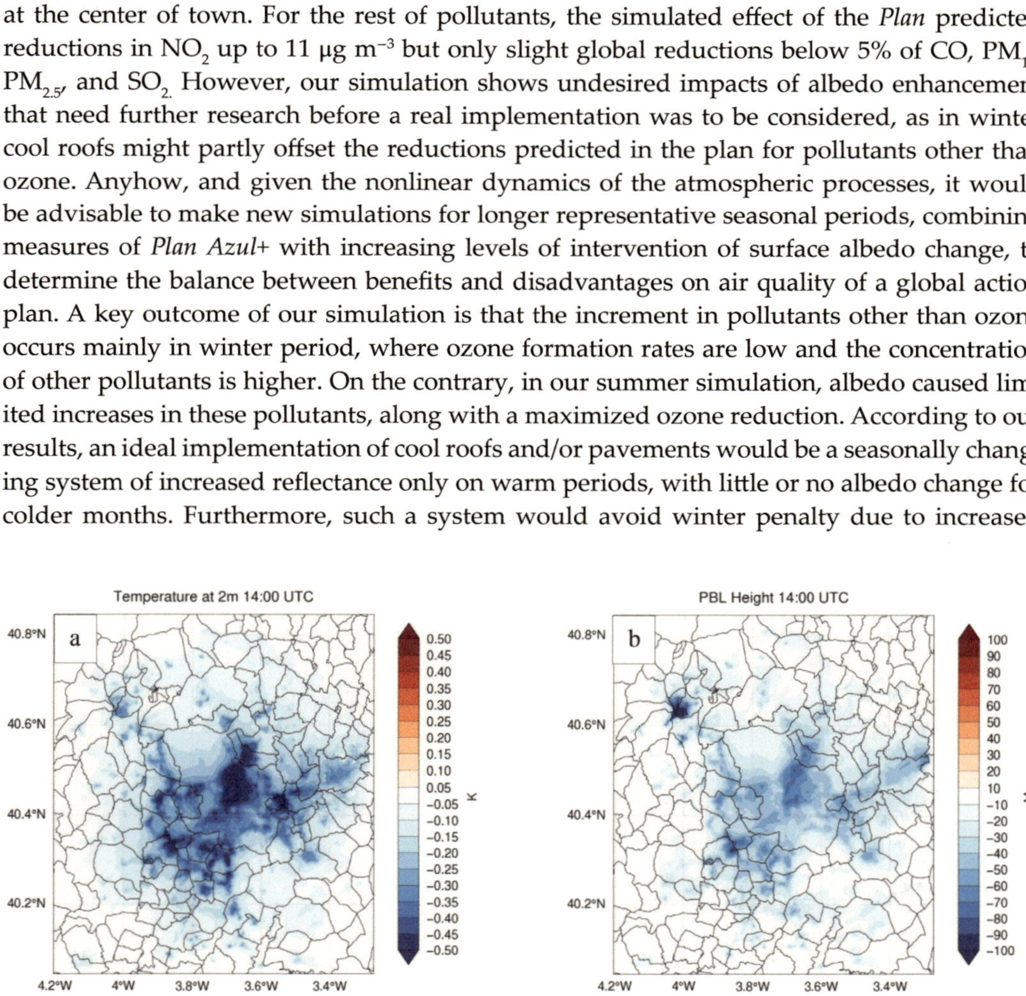

Figure 7. Temperature at 2m (a) and PBL height (b) differences between Alb2 scenario and default scenario for the winter period at 14 UTC.

heating demand [27]. A real experience of seasonal surface albedo change is applied over 20,000 reflective greenhouses in Almeria province, 500 km south of Madrid, where whitewash slaked/lime painting is applied over the roofs to limit excess heating inside the greenhouses in summer and is washed out in September to allow enough winter radiation inside them. The implementation of high albedo in the area has caused mean outdoor surface air temperature cooling, locally offsetting the impact of global warming [28, 29]. The levels of albedo enhancement simulated here (round +0.1 at the pixel level) are similar to those implemented on the field over Almeria area, but with more than double intervention surface at Madrid urban area, well above the minimum critical intervention area for efficient cooling at similar comparable latitudes and insolation. As our temperature data show, expected changes in net solar income at the surface should be comparable in both observed and simulated experiences, with differences in air temperature impact due mostly to location and surface canopy parameters of the urban fabric. However, our results are site and time dependent and have been generated from the specific coupled modeling configuration applied over Madrid city, for the periods simulated, and for the emissions inventory used. Further extensive research with optimized mesoscale modeling should also include the impact of albedo enhancement on cloud cover and precipitation pattern, as rain causes wet deposition of pollutants improving air quality. As our results show, depression of PBL height and dynamics of cooler air over the city might cause reduced vertical mixing and affect to the dilution of pollutants [30]. Other undesirable effects on the microclimate of the city and surrounded areas cannot be discarded and should be studied, such as modifications in the wind pattern and the hydrological cycle in the region [31]. Finally, these results do not account for additional benefits such as reduced cooling energy use and associated reductions in emissions from point sources, neither on the potential negative impacts on heating energy use in winter [3].

Acknowledgements

This study has been supported by the Spanish Government, Ministry of Economy and Competitiveness, Grant No. CGL2013-46873-R. The author is grateful to the Environmental Agency of the Regional Government of Madrid for providing emissions inventory.

Conflict of interest

The author declares no competing financial interest.

Author details

Pablo Campra

Address all correspondence to: pcampra@ual.es

High School of Engineering, University of Almeria, Almeria, Spain

References

[1] Betts RA. Biogeophysical impacts of land use on present-day climate: Near-surface temperature change and radiative forcing. Atmospheric Science Letters. 2001;**2**:39-51

[2] Akbari H, Menon S, Rosenfeld A. Global cooling: Increasing world-wide urban albedos to offset CO_2. Climatic Change. 2009;**94**(4):275

[3] Akbari H, Pomerantz M, Taha H. Cool surfaces and shade trees to reduce energy use and improve air quality in urban areas. Solar Energy. 2001;**70**:295-310

[4] Stathopoulou E, Mihalakakou G, Santamouris M, Bagiorgas HS. On the impact of temperature on tropospheric ozone concentration levels in urban environments. Journal of Earth System Science. 2008;**117**:227-236

[5] Taha H, Douglas S, Haney J. Mesoscale meteorological and air quality impacts of increased urban albedo and vegetation. Energy and Buildings. 1997;**25**:169-177

[6] Taha H, Chang S, Akbari H. Meteorological and Air Quality Impacts of Heat Island Mitigation Measures in Three U.S. Cities. Berkeley, CA, USA: LBNL; 2000

[7] Taha H. Urban surface modification as a potential ozone air-quality improvement strategy in California: A mesoscale modeling study. Boundary-Layer Meteorology. 2008;**127**:219-239

[8] Touchaei AG, Akbari H, Tessum CW. Effect of increasing urban albedo on meteorology and air quality of Montreal (Canada)—Episodic simulation of heat wave in 2005. Atmospheric Environment. 2016;**132**:188-206

[9] Fallmann J, Forkel R, Emeis S. Secondary effects of urban heat island mitigation measures on air quality. Atmospheric Environment. 2016;**125**:199-211

[10] Levinson R, Akbari H. Potential benefits of cool roofs on commercial buildings: Conserving energy, saving money, and reducing emission of greenhouse gases and air pollutants. Energy Efficiency. 2009;**3**:53-109

[11] Arasa R, Domingo-Dalmau A, Vargas R. Using a coupled air quality modeling system for the development of an Air Quality Plan in Madrid (Spain): Source apportionment and analysis evaluation of mitigation measures. Journal of Geoscience and Environment Protection. 2016;**4**:46-61

[12] Salamanca F, Krpo A, Martilli A. A new building energy model coupled with an urban canopy parameterization for urban climate simulations—Part I. formulation, verification, and sensitive analysis of the model. Theoretical and Applied Climatology. 2010;**99**:331-344

[13] Yagüe C, Zurita E, Martínez A. Statistical analysis of the Madrid urban heat island. Atmospheric Environment. 1991;**25**(3):327-332

[14] Sobrino JA, Oltra-Carrió R, Sòria G, Jiménez-Muñoz JC, Franch B, Hidalgo V, et al. Evaluation of the surface urban heat island effect in the city of Madrid by thermal remote sensing. International Journal of Remote Sensing. 2013;**34**(9-10):3177-3192

[15] Denby B. Guidance on the use of models for the European air quality directive. A Working Document of the Forum for Air Quality Modelling in Europe FAIRMODE. ETC/ACC Report. Version 6.2; 2010

[16] Skamarock WC, Klemp JB, Dudhia J, Gill DO, Barker DM, Duda MG, Huang XY, Wang W, Powers JG. A description of the Advanced Research WRF version 3. NCAR Technical Note 475; 2008

[17] WUDAPT. The World Urban Database and Access Portal Tools. 2017. Available from: http://geopedia.world/#T4_L107_x-407046.2848756347_y4928506.709771593_s11_b17 [Accessed: July 8, 2017]

[18] See L, Perger C, Duerauer M, et al. Developing a community-based worldwide urban morphology and materials database (WUDAPT) using remote sensing and crowdsourcing for improved urban climate modelling. In: IEEE Conference Joint Urban Remote Sensing Event (JURSE). Lausanne, Switzerland; 2015

[19] Arasa R, Lozano A, Codina B. Evaluating mitigation plans over traffic sector to improve NO_2 levels in Andalusia (Spain) using a regional-local scale photochemical modeling system. Open Journal of Air Pollution. 2014;**3**:70-86

[20] Bieser J, Aulinger A, Matthias V, Quante M, Denier van der Gon H. Vertical emission profiles for Europe based on plume rise calculations. Environmental Pollution. 2011;**159**:2935-2946

[21] Byun DW, JKS C, editors. Science Algorithms of the EPA Models-3 Community Multiscale Air Quality (CMAQ) Modeling System. Washington D.C., USA: Environmental Protection Agency; 1999

[22] Yarwood G, Rao S, Yocke M, Whitten GZ. 2005. Updates to the Carbon Bond Chemical Mechanism: CB05. Final Report prepared for U.S.EPA

[23] Carlton AG, Bhave PV, Napelenok SL, Edney EO, Sarwar G, Pinder RW, et al. Model representation of secondary organic aerosol in CMAQv4.7. Environmental Science and Technology. 2010;**44**:8553-8560

[24] De la Paz D, Borge R, Martilli A. Assessment of a high resolution annual WRF-BEP/CMAQ simulation for the urban area of Madrid (Spain). Atmospheric Environment. 2016;**144**:282-296

[25] Akbari H, Rose LS, Taha H. Analyzing the land cover of an urban environment using high-resolution orthophotos. Landscape and Urban Planning. 2003;**63**:1-14

[26] WHO. Air Quality Guidelines for Europe. World Health Organization Regional Publications; 2000. European Series No. 91

[27] Akbari H, Konopacki S. Calculating energy-saving potentials of heat-island reduction strategies. Energy Policy. 2005;**33**:721-756

[28] Campra P, Garcia M, Canton Y, Palacios-Orueta A. Surface temperature cooling trends and negative radiative forcing due to land use change toward greenhouse farming in Southeastern Spain. Journal of Geophysical Research. 2008;**113**:D18109

[29] Campra P, Millstein D. Mesoscale climatic simulation of surface air temperature cooling by highly reflective greenhouses in SE Spain. Environmental Science and Technology. 2013;**47**(21):12284-12290. DOI: 10.1021/es402093q

[30] Sailor DJ. Simulated urban climate response to modifications in surface albedo and vegetative cover. Journal of Applied Meteorology. 1995;**34**:1694-1704

[31] Georgescu M, Mahalov A, Moustaoui M. Seasonal hydroclimatic impacts of Sun Corridor expansion. Environmental Research Letters. 2012;**7**:034026/1-034026/9